THE EVOLUTION OF UNTETHERED COMMUNICATIONS

Committee on Evolution of Untethered Communications

Computer Science and Telecommunications Board

Commission on Physical Sciences, Mathematics, and Applications

National Research Council

NATIONAL ACADEMY PRESS
Washington, D.C. 1997

NOTICE: The project that is the subject of this report was approved by the Governing Board of the National Research Council, whose members are drawn from the councils of the National Academy of Sciences, the National Academy of Engineering, and the Institute of Medicine. The members of the committee responsible for the report were chosen for their special competences and with regard for appropriate balance.

This report has been reviewed by a group other than the authors according to procedures approved by a Report Review Committee consisting of members of the National Academy of Sciences, the National Academy of Engineering, and the Institute of Medicine.

Support for this project was provided by the Defense Advanced Research Projects Agency. Any opinions, findings, conclusions, or recommendations expressed in this material are those of the authors and do not necessarily reflect the views of the sponsors.

Library of Congress Catalog Card Number 97-80464
International Standard Book Number 0-309-05946-1

Additional copies of this report are available from:
National Academy Press
2101 Constitution Ave., NW
Box 285
Washington, DC 20055
800-624-6242
202-334-3313 (in the Washington metropolitan area)
http://www.nap.edu

Printed in the United States of America

COMMITTEE ON EVOLUTION OF UNTETHERED COMMUNICATIONS

DAVID J. GOODMAN, Rutgers University, *Chair*
NORMAN ABRAMSON, ALOHA Networks, Inc.
EUGENE CACCIAMANI, Hughes Network Systems
JOEL ENGEL, Ameritech
MARK EPSTEIN, QUALCOMM, Inc.
BRUCE FETTE, Motorola, Inc.
DOUGLAS C. FIELDS, United Parcel Service
BEZALEL GAVISH, Vanderbilt University
ANDREA GOLDSMITH, California Institute of Technology
RANDY H. KATZ, University of California at Berkeley
EDWIN A. KELLEY, Hughes Aircraft Company
KAVEH PAHLAVAN, Worcester Polytechnic Institute
CHARLES E. PERKINS, Sun Microsystems
THEODORE RAPPAPORT, Virginia Polytechnic Institute
JESSE RUSSELL, AT&T Laboratories

COMPUTER SCIENCE AND TELECOMMUNICATIONS BOARD

COMMISSION ON PHYSICAL SCIENCES, MATHEMATICS, AND APPLICATIONS

Preface

In 1994, the Defense Advanced Research Projects Agency (DARPA) initiated the Global Mobile Information Systems (GloMo) program to apply advances in high-speed computation, signal processing, and miniaturization to mobile, wireless, multimedia information systems. The GloMo program is intended to develop the technologies that will enable military forces to carry out communication and computing tasks free of tethers—that is, cables to power sources or telecommunications networks. The concept of "untethered" communications unites mobile and wireless operations.

In response to a request from DARPA, the Computer Science and Telecommunications Board (CSTB) of the National Research Council initiated a one-year study on untethered communications in July 1996. To carry out the study, the CSTB appointed a committee of 15 wireless-technology experts, including researchers, program managers, technology developers, and users working in industry and academia. The Committee on the Evolution of Untethered Communications was charged with advising DARPA on where to invest in information technology for mobile wireless systems. In particular, DARPA posed the following questions:

- What will industry do on its own? When?
- What are the military requirements and needs in mobile wireless information technology?
- To what extent will commercial technology development support military needs?

• Where can DARPA have the greatest impact in technology development or acceleration of technology development to support the military?

This report presents the results of the CSTB study. In addition to answering DARPA's questions, the report provides a wealth of information of interest to a broader audience, suggesting that this is an era of unprecedented change, growth, and promise in untethered communications for both civilian and military users. The report explores the evolution of wireless technology, the often-fruitful synergy between commercial and military research and development efforts, and the technical challenges still to be overcome. Many examples of past ingenuity and future opportunities in wireless systems are noted. Although much of the information can be obtained piecemeal from other sources, it is rarely collected and analyzed in the manner found here. This is the CSTB's first report on wireless communications.

The committee met four times during the one-year study time frame. The report is based on the committee's discussions with representatives of military organizations and commercial developers, background information from the literature, the expertise and judgment of individual committee members, and the deliberations of the committee as a group.

The committee appreciates DARPA's sponsorship of this project and in particular acknowledges the participation of four individuals. Howard Frank, former director of the Information Technology Office, provided valuable advice to the CSTB prior to the initiation of the study. Barry Leiner, the original driving force behind the GloMo program, had the vision to appreciate how a comprehensive technology assessment could complement the specific research projects already under way. Kevin Mills, who later assumed leadership of the GloMo program, addressed the committee at the beginning of its study. Rob Ruth, who succeeded Mills, shared his insights into operational military needs and encouraged the committee and CSTB staff to support DARPA's needs to advance and refine GloMo planning. All of these individuals provided regular reminders of DARPA's strong interest in this study and helped motivate the committee's efforts to develop a comprehensive analysis that takes into account the context of military decision making.

The committee also benefited enormously from the assistance of a number of anonymous reviewers. Reviewer comments and constructive criticisms helped the committee transform a voluminous early draft into a tighter, well-organized final report, calibrate the emphasis placed on a variety of topics, and clarify the analysis of specific issues.

Finally, the committee appreciates the effort and energy devoted to this project by former CSTB staff member Paul Semenza, who organized the study and guided the writing and revision of this report, and former

CSTB project assistant Gail Pritchard, who provided administrative and logistical support. It also is grateful for the yeoman efforts of Laura Ost, the writer-editor who worked with the committee during the summer of 1997 to improve the organization and written presentation of its ideas.

David Goodman, *Chair*
Committee on Evolution of
Untethered Communications

Contents

Executive Summary

Extraordinary growth is evident in capabilities for "untethered" communications, the union of wireless and mobile technologies. An example is cellular telephones, which were virtually unheard of in 1980 and are now used by almost 200 million subscribers worldwide. The subscriber base for wireless communications services is growing 15 times faster than the subscriber base for wired services; by 2010 wireless and wired systems are expected to serve equal numbers of users. The vigorous public demand for wireless services is fueling intense industrial and government activity, including research and development (R&D) aimed at improving the quality and reducing costs of wireless technology, design of innovative systems and services, and implementation of new technical standards and policies. This dynamic environment is producing diverse wireless technologies and standards, in stark contrast to other areas of communications marked by a convergence toward uniformity. All this activity will bring the reality of the next century close to the vision of "anytime, anywhere" communications.

Historically, U.S. national defense needs have stimulated many advances in wireless communications technologies, and the Department of Defense (DOD) has been among the first users. Today, however, some DOD systems lag the state of the art. Whereas the commercial sector has greater incentives than ever before to push the technology envelope—at a cost driven down by a growing mass market—some military wireless equipment is based on 1970s and 1980s technology. The adequacy of the current defense communications infrastructure was called into question during the Gulf War, when voice and data systems failed to keep up with rapidly moving front-line troops. Only 10 percent of soldiers currently have voice communications capabilities, and only satellites, certain aircraft, and "smart" missiles carry sensors for still imagery or video.

Changes in military operations are stimulating DOD interest in untethered systems. U.S. military personnel now need to be prepared to move quickly throughout the world to respond to rapidly evolving regional conflicts and carry out a variety of noncombat roles, such as peacekeeping and humanitarian response. Just as past hardware advances (e.g., aircraft carriers, long-range jet aircraft) shaped the military conflicts of yesteryear, information technology is now shaping plans for the nation's future defense. Plans are being made for a digitized battlefield in which sensors are widely distributed, and rapidly deployable, multimedia wireless systems extend from front-line soldiers all the way to the Pentagon and the North Atlantic Treaty Organization (NATO). Advanced command, control, communications, computing, and intelligence (C^4I) systems will make it possible to monitor an adversary on a computer screen, target specific threats, and neutralize them with the press of a button.

The DOD is taking a dual approach to meeting its future communications needs by funding selected R&D and demonstration projects, focusing primarily on components, while also relying increasingly on commercial off-the-shelf (COTS) technologies. In the Gulf War, for example, the military obtained satellite-transmitted positioning data using commercial receivers, which were rapidly fielded to meet an urgent military need. On the other hand, for some military applications, commercial products do not meet stringent requirements for security, interoperability, and other capabilities. And yet, with defense budgets flat or declining, the DOD can no longer rely solely on military suppliers to provide defense-unique solutions. The military needs to find a way to ride the wave of commercial technology advances while maintaining technical capabilities that exceed those of any potential adversary.

This report, the result of a one-year study by the Computer Science and Telecommunications Board (CSTB) of the National Research Council, recommends strategies and R&D to help the DOD field state-of-the-art, cost-effective untethered communications systems that meet military needs. The report concentrates on wireless technologies that use the radio frequency (RF) part of the electromagnetic spectrum. The study was funded by the Defense Advanced Research Projects Agency (DARPA) to address the following questions:

- What technologies, products, and services will the commercial industry make available on its own, and when?

- What are the military requirements and needs in mobile wireless information technology?
- To what extent will commercial technology development support military needs?
- Where can DARPA have the greatest impact in technology development or acceleration of technology development to support the military?

Answers to those questions are outlined in this summary, which contains two major sections. The first section presents the report's five conclusions and summarizes the supporting facts and analysis contained in Chapters 1 through 3 of the report. The conclusions represent the expert judgments of the 15 members of the CSTB's Committee on the Evolution of Untethered Communications. The second section presents the committee's 12 recommendations, which are based on the conclusions and the supporting facts and analysis. The recommendations, which are directed to DOD and DARPA, are discussed in detail in Chapter 4. It should be noted that although the report focuses on military needs, it also contains a wealth of information about commercial technology developments as well as a primer on the many technical challenges involved in designing wireless systems. The report therefore should also interest a wider, civilian audience.

CONCLUSIONS

A large gap remains between public expectations for mobile communications and the available technology. Voracious consumer demand is stimulating many advances in wireless communications technology, particularly cellular and cordless telephones. As of 1997 there were more than 50 million cellular subscribers in the United States. The portfolio of wireless services now available in the commercial marketplace includes a wide range of telephony, paging, and data applications delivered over a variety of service offerings, ranging from land-based mobile radio to cellular and satellite communications. Each service offers a unique combination of coverage region, bandwidth (i.e., capacity), subscriber equipment features, and connectivity.

In the aggregate, commercial wireless capabilities are considerable. Yet many technical challenges remain. Wireless systems, especially those serving mobile users, are extremely complex. A network needs to be capable of rerouting information seamlessly and efficiently as users move, and sophisticated digital signal processors (DSPs) and antennas are needed to minimize interference, distortion, jamming, and interception without undue power burdens on portable devices. The cost of wireless voice systems remains high compared to that of wired networks, and transmission quality and security could be improved. Specialized wireless data networks have not taken off as yet, perhaps because they are not powerful enough or because two-way mass market applications have yet to emerge.

For a worldwide operator, the management and coordination of diverse systems are complicated by the absence of any trend toward convergence toward a single standard in wireless communications. One digital wireless technology, global system for mobile communications (GSM), is deployed throughout Europe and in more than 100 countries worldwide, whereas the Japanese use their own technology, the personal handyphone system (PHS), and the United States supports three competing technologies: GSM, a time-division system (IS-136), and a code-division system (IS-95). The diversity of technologies in the United States is a result of spectrum regulation policies, which require only that systems not interfere with one another. These policies leave interoperability and other system properties such as quality and efficiency to be settled in the marketplace.

Over the next 10 years or so, market forces will fill the gap between public expectations and the available technology by developing new technologies for commercial wireless communications. Fueled by the success of cellular communications and projections of ever-expanding markets for wireless services, U.S. and foreign industries are performing extensive R&D to overcome remaining technical challenges in wireless systems. For example, efforts are under way to enable portable devices to communicate at the high bit rates needed for advanced information services. In addition, to foster economies of scale in R&D and manufacturing in a world of diverse and changing technical standards, considerable effort is dedicated to advancing the technology of software radios. These radios, by downloading different types of operating software, can serve as single platforms that transmit signals conforming to a variety of standards.

The European Union, which supports cooperative research leading to development of precompetitive technologies, has a vision of the future extending beyond 2002, when universal mobile telecommunication services are scheduled to be deployed. That vision and various industry road maps suggest that, by early in the twenty-first century, commercial wireless communications will achieve the "anytime, anywhere" paradigm.

This optimism does not extend to prospects for fully interoperable wireless communications systems anytime soon. In the United States, wireless communications research is performed by individual companies in the context of their own product plans, and coordination occurs primarily in standards-setting organizations (e.g., Telecommunications Industry Association, Internet Engineering Task Force). Moreover, in the past there was a proliferation of proprietary rather than open network interfaces. The trends toward open systems and digital components will make it easier to build customized systems, but the capabilities of commercial services will continue to depend on consumer demand as well as trade-offs between technology availability and costs.

The military has much to gain from positioning itself to use COTS communications equipment to the greatest extent possible. The civilian and military sectors have a long history of interaction in the design and deployment of wireless technology, such as mobile radios and satellite systems. Commercial wireless technologies are now more attractive than ever in terms of their performance, quality, and cost. By acquiring commercial equipment when it meets military needs, the DOD can field state-of-the-art equipment while also lowering its costs by benefiting from the economies of scale achievable in mass-market manufacturing. For example, software radios for military applications can be built using many COTS components, such as analog-to-digital converters, DSPs, RF amplifiers, displays, batteries, and data-storage devices.

The insertion of commercial technologies into military systems is not always easy. As an example, asynchronous transfer mode (ATM) offers many attractive features, such as high-speed transmission, fast switching, the capability to assign message priority, and queue management. It could provide the basis for improved situational awareness, enabling the DOD to provide real-time imagery, cryptographic security, and low-cost devices for wide distribution. However, the integration of ATM and Internet protocols into wireless battlefield communications will require sophisticated link protocols. Industry is addressing these issues, but the DOD can ensure that its needs are met only by participating in standards-setting activities to influence technology directions and by testing emerging COTS products in battlefield exercises.

Some military needs for wireless communications technologies will exceed or differ significantly from anticipated commercial developments. For example, the military has unique concerns with respect to network design, security, interoperability, and multimode/multiband systems. Although there is clearly overlap between the capabilities of commercial technologies and the DOD's needs, they also differ in a number of respects (see Table ES-1). For example, commercial R&D on integrated (i.e., multimedia) systems is oriented toward base-station-oriented network architectures, an efficient and reliable design in which mobile users communicate with central access nodes. An alternative is the easily reconfigurable peer-to-peer architecture, in which network elements communicate directly. It is not yet clear which architecture will be the most appropriate in future military settings. Improved modeling and simulation tools, especially for communications traffic and mobility of network

TABLE ES-1 Comparison of Commercial Wireless Technologies and Future Military Needs

Feature	Commercial Systems	Future Military Needs
Architecture	Base-station oriented	Not clear at this time
Mobility	Mobile terminals, fixed infrastructure	Mobile terminals, many mobile infrastructure elements
Deployment strategy	Site-specific planning and measurements	Rapid deployment at unpredictable locations
Modes/waveforms	Six at most	Currently 1 to 20; ideal is multimode systems using adaptive waveforms
Frequency range (per system)	Narrowband (one band per system)	Multiband (e.g., 2 megahertz to 2 gigahertz)
Data rate	384 kilobits per second to 2 megabits per second (Mbps) by 2002	Long-term goal is at least 10 Mbps
System access	Anyone who can pay	Universal
Interoperability of systems	Increasingly important	Required for all defense networks and foreign allies
Security	Increasingly important	Critical
Antijam	May become an issue with widespread system use	Critical
Low probability of detection/ interception	Not an issue at this time	Critical
Robustness of systems and equipment	Yes, under moderate conditions	Yes, under extreme conditions (e.g., extreme temperatures, shock, vibrations, submersion)
Interference rejection	Somewhat important	Critical

elements, would support realistic analyses of complex military networks and the design of appropriate protocols and optimization algorithms.

Commercial capabilities and military needs also differ with respect to the security of networks, radio links, and hardware. Commercial users are concerned mainly with privacy and the prevention of unauthorized access to their hardware and data. Some security breaches are tolerated (in fact, analog cellular telephones typically provide no link security). By contrast, the military requires end-to-end encryption to prevent unauthorized access and monitoring of network activity. The military also requires hardware security to prevent an adversary who is opening a device from discerning the hardware or software secrets. Military systems also benefit from antennas and other technologies designed to make signals difficult to detect, jam, or intercept.

Interoperability is another area in which the DOD's needs exceed the interests of the commercial sector. Advanced military wireless systems need to be compatible with the 17 legacy communications networks as well as systems operated by NATO and the United Nations. One means of achieving this objective is the software radio. Now the focus of several military R&D programs, the ideal military software radio is a multimode, multiband unit using many different waveforms over a broad frequency range, whereas commercial versions are likely to offer less flexibility and operate in a single frequency band. To make optimal use of this promising technology, the DOD needs to support specialized R&D focusing on antennas, filters, and adaptive waveforms.

The commercial sector has its own incentives to produce advanced communications devices, components, and subsystems as well as complete systems. To use commercial technologies effectively, the DOD will have to take special measures to promote the development and acquisition of COTS products that can be integrated into systems that meet specialized military requirements. Current commercial R&D and standards activities seek to enable the transmission of many types of information, including data, video, and images, to and from portable wireless devices. Although this work is certain to create new technology, the commercial deployment of the technology is not assured. The availability of the technology in the marketplace will depend on business, social, and government policy factors. Military planners will need to maintain a continuing awareness of the differences between what is possible technically and what is available in the market to meet military needs.

As part of this process, the DOD needs to translate its operational requirements into technical specifications that can be used to determine the suitability of commercial wireless technologies for military applications. New approaches to procurement, as well as technology demonstration and testing rather than development, will help DOD obtain the greatest return on its investments. In addition, by fully understanding the barriers to synergy between the commercial and defense sectors, the DOD can develop processes for accommodating or overcoming these barriers.

For example, the commercial sector tends to add functions to equipment only if economically justified by customer demand; by participating in standards-setting activities the DOD can encourage the design of COTS products that can be more easily modified to meet military needs.

RECOMMENDATIONS

The committee's 12 recommendations are organized from the general to the specific. The first three recommendations identify organizational changes that DOD should make to foster an environment conducive to the absorption of state-of-the-art commercial wireless communications technologies. The other nine recommendations identify research that should be undertaken by DARPA to fill gaps in commercial development efforts and ensure that advanced wireless systems meet military needs. The research recommendations are presented in order of priority, reflecting the committee's view that top-level systems issues are paramount. The remaining recommendations deal with subsystems and components.

Organizational Changes

1. The DOD should participate in standards-setting activities for wireless communications technologies and systems.

2. The DOD should pursue a vigorous process of technology demonstration and testing prior to extensive development and procurement. In particular, the focus should be on systems concepts based on commercial technologies and specialized military enhancements.

3. The DOD should plan a new approach to procurement that will identify how commercial infrastructure systems and subscriber equipment can best be used for military purposes and how to purchase commercial equipment in the most productive way.

Systems Research

4. DARPA should build on current research in modeling and simulation to incorporate the communications traffic, mobility of network elements, and radio propagation encountered in mobile military information networks.

5. DARPA should initiate research to produce network architectures that incorporate commercial products in a manner that meets military requirements.

6. DARPA should conduct research aimed at understanding and bridging the differences between security needs in commercial and military networks.

7. DARPA should conduct research aimed at reducing co-site interference.

Component Research

8. DARPA should carry out research and demonstration projects designed to field software radio technology for military applications.

9. DARPA should conduct the research needed to adapt smart antennas for mobile military applications.

10. DARPA should conduct research to produce transmission techniques that adapt to a wide range of operating conditions.

11. DARPA should conduct research to overcome the limitations of current filter technology for use in military software radios and high-density platforms.

12. DARPA should develop novel components to enhance the flexibility of software radios.

1

Past, Present, and Future

Humans have long dreamed of possessing the capability to communicate with each other anytime, anywhere. Kings, nation-states, military forces, and business cartels have sought more and better ways to acquire timely information of strategic or economic value from across the globe. Travelers have often been willing to pay premiums to communicate with family and friends back home. As the twenty-first century approaches, technical capabilities have become so sophisticated that stationary telephones, facsimile (fax) machines, computers, and other communications devices—connected by wires to power sources and telecommunications networks—are almost ubiquitous in many industrialized countries. The dream is close to becoming reality. The last major challenge is to develop affordable, reliable, widespread capabilities for "untethered" communications, a term coined by the U.S. military and referring to the union of wireless and mobile technologies. Because "untethered" is not a widely used term, this report concentrates on "wireless" communications systems that use the radio frequency (RF) part of the electromagnetic spectrum. These systems and their component technologies are widely deployed to serve mobile users.

Mobile wireless communications is a shared goal of both the U.S. military and civilian sectors, which traditionally have enjoyed a synergistic relationship in the development and deployment of communications technology. The balance of that long-standing interdependence is changing now as a result of trends in the marketplace and defense operations and budgets. These trends suggest that market forces will propel advances in technology to meet rising consumer expectations. However, the military may need to take special measures to field cost-effective, state-of-the-art untethered communications systems that meet defense requirements.

This chapter lays the foundation for an analysis of military needs in this area by chronicling the evolution of military and civilian applications of communications technology, from ancient times leading up to the horizon of 2010. Section 1.1 is an overview of the challenge facing the U.S. military. Section 1.2 provides an historical perspective on the development of communications infrastructures. Section 1.3 outlines the wireless systems currently used by the U.S. military and the related research and development (R&D) activities. Sections 1.4 through 1.7 recount the evolution and current status of commercial wireless systems. Section 1.8 compares the development paths for wireless technologies in the United States, Europe, and Japan.

1.1 OVERVIEW

In the final years of the twentieth century, all aspects of wireless communications are subject to rapid change throughout the world. Dimensions of change include the following:

- Vigorously expanding public demand for products and services;
- Dramatic changes worldwide in government policies regarding industry structure and spectrum management;
- Rapidly advancing technologies in an atmosphere of uncertainty about the relative merits of competing approaches;
- Emergence of a wide variety of new systems for delivering communications services to wireless terminals; and
- Profound changes in communications industries as evidenced by an array of mergers, alliances, and spin-offs involving some of the world's largest corporations.

These changes are fueled by opportunities for profit and public benefit as perceived by executives, investors, and governments. Although the patterns are global, the details differ significantly from country to country. Each dimension of change is complex and all of them interact. Overall, the dynamic nature of wireless communications creates a mixture of confusion and opportunity for stakeholders throughout the world.

A principal attraction of wireless communications is its capability to serve mobile users. Because mobility is an important feature of military operations, the U.S. armed forces have always played a leading role in the development and deployment of wireless communications technology.

In the coming years, however, it appears that the commercial sector will have sufficient incentives and momentum to push the technical envelope on its own. At the same time, flat or declining defense budgets are motivating the military to adopt commercial products and services to an increasing extent. Yet there are significant differences between military and commercial requirements. Thus, it is important to examine carefully the opportunities for, and limitations to, military use of commercial wireless communications products and services.

In contrast to other areas of information technology, wireless communications has yet to converge toward a single technical standard or even a very small number of them. Instead it appears that diversity will endure for the foreseeable future. In this environment, the management and coordination of complex, diverse systems will be an ongoing challenge, particularly for the U.S. military, which coincidentally has to adapt to new threats and responsibilities after more than half a century of following the paradigm set by World War II and the Cold War. Information is now assuming greater strategic importance than ever before in warfare and other military operations, and so the wide deployment of cost-effective, state-of-the-art wireless communications systems has become particularly critical.

The present situation recalls previous epochs in which breakthroughs in hardware—aircraft carriers, jet aircraft, tactical missiles, nuclear weapons—have led to radical revisions of military doctrine. The next great revolution in military affairs could be shaped by information technology: global communications, ubiquitous sensors, precision location, and pervasive information processing. Advanced command, control, communications, computing, and intelligence (C^4I) systems could make it possible to monitor an adversary, target specific threats, and neutralize them with the best available weapon. Admiral William Owens, former vice chairman of the Joint Chiefs of Staff, has called such an integrated capability a "system of systems." Using such a system, a commander could observe the battle from a computer screen, select the most threatening targets, and destroy them with the press of a button. Battles would be won by the side with the best information, not necessarily the one with the largest battalions.

But unlike the military hardware of the past, information technology is advancing at a breakneck pace in a worldwide marketplace, driven not by military requirements but by the industrial and consumer sectors. Increasingly these technologies are available worldwide, and the best technology is no longer limited to U.S. manufacture and control. Highly accurate position data transmitted by satellite are now available to any yachtsman. High-resolution satellite photographs are for sale around the world. Any nation can purchase the latest communications gadgets from the electronics stores of Tokyo.

Therein lies the challenge for the U.S. military: how to exploit the advances in affordable technology fueled by worldwide consumer demand while also maintaining technical capabilities that significantly exceed those of any potential adversary.

1.2 HISTORICAL PERSPECTIVE

Throughout most of history, the evolution of communications technologies has been intimately intertwined with military needs and applications. Some of the earliest government-sponsored R&D projects focused on communications technologies that enabled command and control. A synergistic relationship then evolved between the military and commercial sectors that accelerated the technology development process. Now large corporations develop the latest communications technologies for international industrial and consumer markets shaped by government regulation and international agreements. World trade in telecommunications equipment and services was valued at $115 billion in 1996 (*The Economist*, 1997).

Modern wireless communication systems are rooted in telephony and radio technologies dating back to the end of the nineteenth century and the older telegraphy systems dating back to the eighteenth century. Wireless systems are also influenced by and increasingly linked to much newer communications capabilities, such as the Internet, which originated in the 1960s. All wireless systems transmit signals over the air using different frequency transmission bands designated by government regulation. Table 1-1 provides an overview of wireless RF communications systems and services and the frequency bands they use.[1] Each frequency band has both advantages and disadvantages. At low frequencies the signal propagates along the ground; attenuation is low but atmospheric noise levels are high. Low frequencies cannot carry enough information for video services. At higher frequencies there is less atmospheric noise but more attenuation, and a clear line of sight is needed between the transmitter and receiver because the signals cannot penetrate objects. These frequencies offer greater bandwidth, or channel capacity.

1.2.1 Communications Before the Industrial Age

The annals of antiquity offer examples of muscle-powered communications: human runners, homing pigeons, and horse relays. Perhaps the earliest communications infrastructure was the road network of Rome, which carried not only the legions needed to enforce the emperor's will

TABLE 1-1 Overview of Wireless Radio Frequency Communications Systems and Services

Frequency Band[a]	Communications Applications	Characteristics
3–30 kHz (very low, or VLF) 30–300 kHz (low, or LF)	Long-range navigation, marine radio beacons	Low attenuation, high atmospheric noise
300–3000 kHz (medium, or MF) 3–30 MHz (high, or HF)	Maritime radio, AM radio, telephone, telegraph, facsimile	Attenuation varies, noise drops at 30 MHz
30–300 MHz (very high, or VHF) 0.3–3 GHz (ultrahigh, or UHF)	VHF television, FM two-way radio, UHF television, radar	Cosmic noise, line-of-sight propagation
3–30 GHz (superhigh, or SHF)	Satellite, radar, microwave	Atmospheric attenuation
30–300 GHz (extremely high, or EHF)	Experimental satellite, radar	Line-of-sight propagation

[a]Frequencies are in kilohertz (kHz), megahertz (MHz), and gigahertz (GHz).

SOURCE: Adapted from Couch (1995).

but also messengers to direct forces far from the capital. Ancient societies also developed systems that obviated the need for physical delivery of information. These systems operated within line-of-sight distances (later extended by telescope): smoke signals, torch signaling, flashing mirrors, signal flares, and semaphore flags (Holzman and Pehrson, 1995). Observation stations were established along hilltops or roads to relay messages across great distances.

1.2.2 Telegraphy

The first comprehensive infrastructure for transmitting messages faster than the fastest form of transportation was the optical telegraph, developed in 1793. Napoleon considered this his secret weapon because it brought him news in Paris and allowed him to control his armies beyond the borders of France. The optical telegraph consisted of a set of articulated arms that encoded hundreds of symbols in defined positions. Under a military contract, the signaling stations were deployed on strategic hilltops throughout France, linking Paris to its frontiers. By the mid-1800s, 556 stations enabled transmissions across more than 5,000 kilometers (km).

The optical telegraph was superseded by the electrical telegraph in 1838, when Samuel Morse developed his dot-and-dash code. Now information could be transmitted beyond visible distances without significant delay. In an 1844 demonstration on a government-funded research testbed, Morse sent the message "What Hath God Wrought?" from Baltimore to the U.S. Capitol (Bray, 1995).

The rapid deployment of telegraphic lines around the world was driven by the need of nineteenth-century European powers to communicate with their colonial possessions. High-risk technology investments were required. After the use of rubber coating was demonstrated on cables deployed across the Rhine River, the first transatlantic cable was laid in 1858, but it failed within months. A new cable designed by Lord Kelvin was laid in 1866 and operated successfully on a continuous basis.

The result was a rapidly expanding telegraphic network that reached every corner of the globe. By 1870, Great Britain communicated directly with North America, Europe, the Middle East, and India. Other nations scrambled to duplicate that system's global reach, for no nation could trust its critical command messages to the telegraphic lines of a foreign power.

1.2.3 Early Wireless

Within a few decades of its widespread deployment, telegraphy began to lose customers to a new technology—radio. In 1895 Guglielmo

Marconi demonstrated that electromagnetic radiation could be detected at a distance. Great Britain's Royal Navy was an early and enthusiastic customer of the company that Marconi created to develop radio communications. In 1901 Marconi bridged the Atlantic Ocean by radio, and regular commercial service was initiated in 1907 (Masini, 1996).

The importance of this new technology became evident with the onset of World War I. Soon after hostilities began, the British cut Germany's overseas telegraphic cables and destroyed its radio stations. Then Germany cut Britain's overland cables to India and those crossing the Baltic to Russia. Britain enlisted Marconi to put together a string of radio stations quickly to reestablish communications with its overseas possessions.

The original Marconi radios were soon replaced by more advanced equipment that exploited the vacuum tube's capability to amplify signals and operate at higher frequencies than did older systems. In 1915 the first wireless voice transmission between New York and San Francisco signaled the beginning of the convergence of radio and telephony. The first commercial radio broadcast followed in 1920 (Lewis, 1993). The use of higher frequencies (called shortwaves) exploited the ionosphere as a reflector, greatly increasing the range of communications. By World War II, shortwave radio had developed to the point where small radio sets could be installed in trucks or jeeps or carried by a single soldier. The first portable two-way radio, the Handie-Talkie, appeared in 1940. Two-way mobile communications on a large scale revolutionized warfare, allowing for mobile operations coordinated over large areas.

1.2.4 Telephony

The telephone was first demonstrated in 1876. A telephone network based on mechanical switches and copper wires then grew rapidly. The high cost of the cables limited the number of conversations possible at any one time; as demand increased, multiplexing techniques, such as time division and frequency division, were developed.

A mix of independent operators ran telephone services in the early days. Subscribers to different services could not call each other even when in the same town. In 1913 the U.S. government allowed American Telephone and Telegraph (AT&T) to assume control of the national telephone network in return for becoming a regulated monopoly delivering "universal" service. Yet it was not until the 1950s that unified network signaling was offered to subscribers, allowing them to make direct-dial long-distance telephone calls (Calhoun, 1992). Since then, the rapid extension of the long-distance telephone network has been made possible by advances in photonic communications and network control technologies.

1.2.5 Communications Satellites

The concept of using geosynchronous satellites for communications purposes was first suggested in 1945 by the science fiction writer Arthur C. Clarke, then employed at Britain's Royal Aircraft Establishment, part of the Ministry of Defence. Satellites of this type are positioned above the equator and move in synch with Earth's rotation. In 1954 J.R. Pierce at AT&T's Bell Telephone Laboratories developed the concept of orbital radio relays and identified the key design issues for satellites: passive versus active transmission, station keeping, attitude control, and remote vehicle control (Bray, 1995). Pierce advocated an approach of reaching geostationary orbit in successive stages of technology development, starting with nonsynchronous, low-orbit satellites. Hughes Aircraft Company advocated a geostationary concept based on the company's patented station-keeping techniques.

In 1957 the Soviet Union launched Sputnik, the first satellite to be placed in orbit. Amateur radio operators were able to pick up its low-power transmissions all over the world. In 1960 the National Aeronautics and Space Administration (NASA) and Bell Laboratories launched the first U.S. communications satellite, Echo-1, in a low Earth orbit. The first satellite-based voice message was sent by President Dwight Eisenhower using passive transmission techniques. The next advance in satellite technology was the successful launch of the TELSTAR system by NASA and Bell Laboratories. Using active transmission technology TELSTAR delivered the first television transmission across the Atlantic in 1962. Because it was placed in an elliptical orbit that varied from low to medium altitudes, the satellite was visible contemporaneously to Earth stations on both sides of the Atlantic for only about 30 minutes at a time. Clearly geostationary orbits were desirable if satellites were to be used for continuous telephone and television communications across long distances.

In 1963 Hughes Aircraft and NASA achieved geosynchronous orbit (known as GEO today) with the successful launch of the SYNCOM satellite. The satellite was placed in an orbit of approximately 36,210 km, a distance that allowed it to remain stationary over a given point on Earth's surface. SYNCOM led the way for the next several decades of satellite systems by demonstrating that synchronous orbit was achievable, and that station keeping and attitude control were feasible. Today most satellites, both military and commercial, are of the GEO variety.

COMSAT was formed by an act of Congress in 1962 and represented U.S. commercial interests in satellite technology development at Intelsat, established in 1964 as an international, government-chartered organization to coordinate worldwide satellite communications issues. INTELSAT-II (Early Bird) was launched into a geosynchronous orbit in

1965 and supported 240 telephone links or one television channel. Channel capacities are now measured in the tens of thousands of voice channels (the INTELSAT-VI, launched in 1987, supports 80,000 voice channels).

The first military satellites, the DSCS-I group, were launched by the U.S. Air Force in 1966. Three launches placed 26 lightweight (100-pound) satellites in near-geosynchronous orbit. These systems supported digital voice and data communications using spread-spectrum technology (an important signal-processing approach discussed extensively in Chapter 2). The satellites were replaced in the 1970s by the DSCS-II group, which increased channel capacity by using spot-beam antennas with high gain to boost the received power. The first cross-linked military satellites, the LES 8/9, were launched in 1976. This demonstration fostered a vision of space-based architectures—without vulnerable ground relays—for communication, navigation, surveillance, and reconnaissance.

Satellites offer several advantages over land-based communications systems. Rapid, two-way communications can be established over wide areas with only a single relay in space, and global coverage with only a few relay hops. Earth stations can now be set up and moved quickly. Furthermore, satellite systems are virtually immune to impairments such as multipath fading (channel impairments are discussed in Chapter 2). But with the rapid deployment of undersea fiber-optic links, the use of satellite channels for telephony has been on the decline. The high capacity of fiber provides for competitive costs, which, combined with low latency, have attracted consumers. The future of the satellite industry depends on the emergence of applications other than fixed telephony channels. A new generation of satellite systems is being deployed to provide mobile telephone services (see Section 1.5).

1.2.6 Mobile Radio and the Origins of Cellular Telephony

The early development of mobile radio was driven by public safety needs. In 1921 Detroit became the first city to experiment with radio-dispatched police cars. However, transmission from vehicles was limited by the difficulty of producing small, low-power transmitters suitable for use in automobiles. Two-way systems were first deployed in Bayonne, New Jersey, in the 1930s. The system operated in "push-to-talk" (i.e., half-duplex) mode; simultaneous transmission and reception, or full-duplex mode, was not possible at the time (Calhoun, 1988).

Frequency modulation (FM), invented in 1935, virtually eliminated background static while reducing the need for high transmission power, thus enabling the development of low-power transmitters and receivers for use in vehicles. World War II stimulated commercial FM manufacturing capacity and the rapid development of mobile radio technology. The need for thousands of portable communicators accelerated advances in system packaging and reliability and reduced costs. In 1946 public mobile telephone service was introduced in 25 cities across the United States. The initial systems used a central transmitter to cover a metropolitan area. The inefficient use of spectrum and the coarseness of the electronic filters severely limited capacity: Thirty years after the introduction of mobile telephone service the New York system could support only 543 users.

A solution to this problem emerged in the 1970s when researchers at Bell Laboratories developed the concept of the cellular telephone system, in which a geographical area is divided into adjacent, non-overlapping, hexagonal-shaped "cells." Each cell has its own transmitter and receiver (called a base station) to communicate with the mobile units in that cell; a mobile switching station coordinates the handoff of mobile units crossing cell boundaries. Throughout the geographical area, portions of the radio spectrum are reused, greatly expanding system capacity but also increasing infrastructure complexity and cost.

In the years following the establishment of the mobile telephone service, AT&T submitted numerous proposals to the Federal Communications Commission (FCC) for a dedicated block of spectrum for mobile communications. Other than allowing experimental systems in Chicago and Washington, D.C., the FCC made no allocations for mobile systems until 1983, when the first commercial cellular system—the advanced mobile phone system (AMPS)—was established in Chicago. Cellular technology became highly successful commercially with the miniaturization of subscriber handsets.

1.2.7 The Internet and Packet Radio

The original concepts underlying the Internet were developed in the mid-1960s at what is now the Defense Advanced Research Projects Agency (DARPA), then known as ARPA. The original application was the ARPANET, which was established in 1969 to provide survivable computer communications networks. The ARPANET relied heavily on packet switching concepts developed in the 1960s at the Massachusetts Institute of Technology, the RAND Corporation, and Great Britain's National Physical Laboratory (Kahn et al., 1978; Hafner and Lyon, 1996; Leiner et al., 1997). This approach was a departure from the circuit-switching systems used in telephone networks (see Box 1-1).

The first ARPANET node was located at the University of California at Los Angeles. Additional nodes were soon established at Stanford Research Institute (now SRI International), the University of California at Santa Barbara, and the University of Utah. The development of a host-to-host protocol,[2] the network control protocol (NCP), followed in 1970,

BOX 1-1
Circuit Switching Versus Packet Switching

Telephone systems are based on a connection-oriented or circuit-switched model in which connections are fixed for the duration of a call. Such systems are inefficient when transmission occurs in short bursts separated by long pauses. Packet switching replaces the centralized switches with distributed routers, each with multiple connections to adjacent routers. Messages are divided into "packets" that are independently routed on a hop-by-hop basis. Such an approach allows messages to be multiplexed over the available paths on a statistically determined basis, gracefully adapting the transmissions to traffic levels and optimizing the use of existing link capacity without pre-allocating link bandwidth.

enabling network users to develop applications. At the same time, the ALOHA Project at the University of Hawaii was investigating packet-switched networks over fixed-site radio links. The ALOHANET began operating in 1970, providing the first demonstration of packet radio access in a data network (Abramson, 1985). The contention protocols used in ALOHANET served as the basis for the "carrier-sense multiple access with collision detection" (CSMA/CD) protocols used in the Ethernet local area network (LAN) developed at Xerox Palo Alto Research Center in 1973. The widespread use of Ethernet LANs to connect personal computers (PCs) and workstations allowed broad access to the Internet, a term that emerged in the late 1970s with the design of the Internet protocol (IP). The need to link wired, packet radio, and satellite networks led to the specifications for the transmission control protocol (TCP), which replaced NCP and shifted the responsibility for transmission from the network to the end hosts, thereby enabling the protocol to operate no matter how unreliable the underlying links.[3]

The development of microprocessors, surface acoustic wave filters, and communications protocols for intelligent management of the shared radio channel contributed to the advancement of packet radio technology in the 1970s. In 1972 ARPA launched the Packet Radio Program, aimed at developing techniques for the mobile battlefield, and SATNet, an experimental satellite network. In 1983 ARPA launched a second-generation packet radio program, Survivable Adaptive Networks, to demonstrate how packet radio networks could be scaled up to encompass much larger numbers of nodes and operate in the harsh environment likely to be encountered on the mobile battlefield.

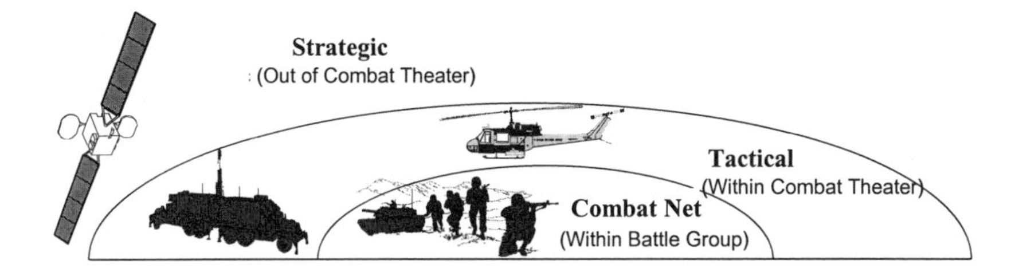

FIGURE 1-1 Military radios are designed for different uses. Combat net radios, for example, are designed for communications within a battle group.

1.3 MILITARY WIRELESS SYSTEMS AND RESEARCH

1.3.1 Terrestrial Systems

Radio communications technology is widely used by U.S. military units at all levels. The many different types of military radios and applications cause a variety of communication problems. The military environment magnifies common difficulties such as the failure of one radio type to communicate with another type (interoperability), failure of one user to communicate with another (connectivity), incompatibility of new radios with old radios (legacy systems), and one radio at a location interfering with another radio at the same location (co-site interference).

In general, U.S. military radio systems can be categorized by the location of users and the information they broadcast and receive (see Figure 1-1). Multiple radios are often gathered together in an aircraft, shipboard radio room, or communications van to form tactical radio complexes and command-and-control centers. The radios operate simultaneously using many different waveforms across several frequency bands (e.g., high frequency [HF], very high frequency [VHF], and ultrahigh frequency [UHF]).

Combat net radios take the form of either a single radio in a vehicle (much like a car radio) or a device like a "walkie-talkie" carried around by a soldier. Most of the information broadcast on combat net radios consists of voice communications, often to share position information. Many of today's combat net radios have been enhanced to carry data in addition to voice. In general, combat net radios have fewer capabilities and cost less than do tactical radios (see Table 1-2). Military radios generally cost much more than commercial systems supporting similar applications.

Deployed military radios have various shortcomings. For example, the mobile subscriber equipment (MSE), the U.S. Army's mobile telephone system for the battlefield, was designed to be like a cellular telephone but is outdated compared to current technology. The single-chan-

TABLE 1-2 Tactical and Combat Net Radios

Characteristics	C^4I Radios[a]	Army Tactical Radios
Simultaneous channels	4–20	1
Waveforms	5–20	1–4
Waveform structure	Wideband and narrowband	Narrowband
Cost	$50,000–$500,000+	$5,000–$50,000
Deployed examples	• Joint tactical information distribution system • Joint Tactical Terminal • Fleet broadcast (UHF-satellite naval command-and-control radio)	• Single-channel ground and airborne radio system • Enhanced position location reporting system • Mobile subscriber equipment

[a]Command, control, communications, computing, and intelligence.

nel ground and airborne radio system (SINCGARS) has been updated with recent technology, including programmable microprocessors, application-specific integrated circuits (ASICs), and surface-mount technology, but it implements a series of outdated waveform standards for single-channel digital voice. Furthermore, SINCGARS has experienced severe co-site interference problems because it hops transmission frequencies within the VHF band, a design capability that helps prevent jamming by adversaries but results in hops onto channels already in use for other communications traffic. The mobile subscriber radio terminal (MSRT) costs $70,000 and is about the size of a microwave oven; an updated version, introduced in 1994, is no less expensive and no smaller. Numerous HF radios have been built by the Army, but most are in storage because these radios are not simple push-to-talk designs and user training for the difficult HF channel has not been widespread.

The problems posed by individual radios are exacerbated by the difficulties encountered in linking communications systems of varying sophistication together (see Box 1-2). Special interfaces can be designed; SINCGARS, for example, can be interfaced into the MSRT. Inherent interoperability is among the features sought in sophisticated future systems. But in the near term, front-line troops will continue to use both existing and evolving radios, such as SINCGARS, mobile tactical satellite (TACSAT) terminals, MSE, MSRT, and packet radios. The Army is struggling with how to upgrade the MSE, a proprietary system. The SINCGARS is expected to be replaced and upgraded with a tri-service joint tactical radio in 1999.

The U.S. Department of Defense established IP as the underlying "building code" for the Army, making a commitment to migrate all communications networks to the same basic structure as the Internet to position the military to integrate and leverage the advances in commercial information technologies. The Army's Task Force XXI "Tactical Internet" (Booz-Allen & Hamilton, 1995) was the first major experimental fielding of this new architecture (Sass and Eldridge, 1994; Sass, 1996).

BOX 1-2
Realities of Military Communications in Bosnia

U.S. military communications systems in Bosnia have been frustrating, according to Brigade Commander Kenneth Allard (1996), who described the situation this way:

> Despite the imperative of supporting the warfighter, the river of information available to U.S. military forces in Bosnia often diminishes to a trickle by the time it reaches the soldiers actually executing peacekeeping missions. On one recent operation, a brigade commander who had requested overhead imagery of his area complained that "the system" took three weeks to provide photographs that eventually turned out to be six months old. The reasons are many: communications pipelines too narrow to efficiently carry digital data to the field, outmoded tactical equipment, and automation resources easily overwhelmed by what data are available.
>
> . . . The Army communications system has generally worked well in Bosnia, but only at great costs in manpower and effort. Because Army tactical radios operate on line-of-sight transmissions, it is essential to place repeaters and relays on mountain tops. But with large numbers of radios nets required for the 15 brigades operating in the U.S. sector, there is a real problem with interference ("signal fratricide"). When these critical relay sites must be fortified and defended, support requirements can consume 7–8 percent of combat manpower in addition to the U.S. signal brigade of over 1,100 soldiers. . . . Although the military communications system features free morale calls, most U.S. soldiers "phone home" with AT&T prepaid credit cards—expense outweighed by clarity and convenience. Their commanders have similar feelings. "The former warring factions have better communications," snapped one U.S. brigade commander, "because they have cellular phones and I don't."

1.3.2 Satellite Systems

Satellite systems play a major role in military communications. They are attractive alternatives to land-based systems because they provide mobile and tactical communications to a large number of users over a wide geographical area. In addition, communication links can be added or deleted quickly, and satellites are less vulnerable to destruction or enemy exploitation than are land-based systems.

The DOD uses both military and commercial satellites to meet its communications needs. Fleet communications are supported by the government-owned FLTSAT and contractor-owned LEASAT systems, both of which are geosynchronous. The U.S. Air Force uses FLTSAT, the elliptical-orbit Satellite Data System, and the DSCS-III satellites to support the AFSATCOM satellite system. The DSCS, a vital component of the global defense communications system, is the DOD's primary system for long-haul, high-volume trunk traffic. The operational DSCS space segment consists of a mix of DSCS-II and DSCS-III satellites.

In 1982 the military began developing new satellite and terminal technology for MILSTAR, a millimeter-wave system operating in the 30–60 gigahertz (GHz) range. This new system consists of both geosynchronous and inclined-orbit satellites. The system provides enhanced antijam (AJ) capabilities as well as hardening against nuclear attack. Only a few of the planned eight MILSTAR satellites have been deployed so far. The complete system would provide two satellites per coverage area over the continental United States and the Atlantic, Pacific, and Indian oceans.

In general, existing tactical-satellite ground terminals incorporate new technology (e.g., microprocessors, ASICs, surface-mount technology) but are still forced to implement legacy waveforms. As a result, they have generally not kept pace with innovations in commercial communications waveforms and standards. In the case of MILSTAR, the military uses a noncommercial frequency band and is therefore unable to use—or take advantage of the price reductions in—commercial hardware. The new Joint Tactical Terminal (one of the systems listed in Table 1-2) is designed using modern radio technology, perhaps even including software-defined radios (see Section 1.3.3.2). High data rates sufficient for multimedia transmissions can be achieved only with the most advanced technology. For example, the global broadcast system (GBS), part of the U.S. Navy's UHF Follow-On satellites 8, 9, and 10, has bandwidth exceeding 100 megabits per second (Mbps) and worldwide coverage.

The most widely used military satellite system is the global positioning system (GPS), which encompasses 18 to 24 satellites in inclined orbits transmitting spread-spectrum signals. The GPS receivers extract precise time and frequency information from these signals to determine with great accuracy the receiver location, velocity, and acceleration. The system can be used by anyone with a receiver.[4] Commercial GPS receivers are used for numerous applications, including surveying, aircraft and ship navigation, and even recreational activities on land. Although launching and upkeep of the entire fleet of satellites are paid for by the United States, commercial GPS receivers were used by both sides in the Gulf War.

1.3.3 Research Initiatives in Untethered Communications

The DOD's vision for future communications systems is typically expressed in general terms, such as "multimedia to the foxhole" (see Box 1-3). For example, the Army's architecture for the digitized battlefield of the twenty-first century consists of fixed high-bandwidth infrastructure at the Army, theater, and corps levels, integrated with the DOD's global grid (a concept for spanning the world with high-bandwidth computing and communications systems) and based on asynchronous transfer mode (ATM) wide-area networking technology (Sass and Gorr, 1995). Bandwidth is allocated not only up and down the command hierarchy but also horizontally to cooperating formations. At the division level and below, wireless extensions provided by mobile radio access points (RAPs) will link the front-line combat communications systems to the infrastructure in the rear areas. The RAP is a wheeled or tracked vehicle with an on-the-move antenna system. The RAPs carry extensive communications systems and are interconnected by high-capacity trunk radios capable of

BOX 1-3
Preparing for Battle in a Multisensor Environment

The Army's Force XXI Soldier Program is developing the prototype technologies needed to make the soldier more efficient as a sensor and more lethal. Many sensors will be used in future theaters of operation: Everything that moves will have one or more sensors, and there will be many stationary sensors. Soldiers may carry position, identification, health, and imagery sensors, for example, or a networking body-worn radio. They will probably be able to image and locate anything on the battlefield and notify others of the onset of a firefight, drawing support from nearby assets. Finally, soldiers will be able to provide reconnaissance reports in far greater detail, perhaps catching important details missed in traditional daily radio reporting. The dismounted soldier is not the only user of imagery and video services; ships, aircraft, and other platforms may also carry sensors. Video and video teleconferencing applications have already been deployed on an experimental basis. All of these sensors will increase the data load on the military communications system.

communicating at up to 45 Mbps over a range of 30 km. Satellites or other systems may provide back-up communications.

To the committee's knowledge, the operational requirements for future untethered communications have not been translated into technical specifications. In the future, technical specifications will need to be formulated in a way that will make it possible to determine which commercial technologies are capable of meeting military needs. As an alternative, some general DOD requirements can be inferred from military plans and the known technical capabilities of existing and emerging communications technologies. For example, future military wireless systems will require high data rates—the long-range goal is at least 10 Mbps—and the capability to transmit over broad and variable frequency bands (some experimental radios are designed to span frequencies from 2 MHz to 2 GHz). The systems will need to be rapidly deployable and the infrastructure will need to be mobile. Multilevel communications security that encompasses the most secure levels possible will be needed. Furthermore, to enable worldwide strategic communications, the new equipment will need to be interoperable with older military systems as well as those used by foreign allies and international forces. There are more than 17 different U.S. defense communications networks, and none are readily interoperable at present. New concepts and technologies will clearly be needed to meet all these requirements.

To meet its future communications requirements, the DOD is funding a number of research and demonstration projects, typically pursuing high-risk ventures with potentially high payoff. The most comprehensive DOD-funded initiative dealing with untethered communications is the Global Mobile Information Systems (GloMo) program initiated by DARPA in 1994. Other relevant research initiatives deal with software-defined radios, communications systems, and radio technology (Leiner et al., 1996).

1.3.3.1 Global Mobile Information Systems Program

The overarching goal of GloMo is to develop technology for robust end-to-end information systems in a global mobile environment by exploiting commercial products and generating new technologies with applications in both commercial and military domains. The program supports a wide range of research projects, which are identified based on the priorities of GloMo managers rather than on a systems approach to the development of top-down solutions. Notably missing from the program, for example, is a comprehensive assessment of the suitability of various network architectures, even though all other component needs are dictated by the system design. (Network architecture issues are discussed in detail in Chapters 2 and 3.) The GloMo program currently focuses on developing innovative technologies that span the following research thrusts.

Design Infrastructure. This effort spans tools, languages, and environments for designing and deploying wireless systems. Research areas include computer-aided design tools for estimating power and designing low-power systems, design libraries and models for mixed-signal integrated circuits (ICs) suitable for implementing highly integrated RF chip sets, and simulation tools for modeling the propagation of radio waves and higher-level protocols.

Untethered Nodes. This effort focuses on high-performance, modular, low-cost, and low-power wireless nodes. Research activities are aimed at developing the next generation of agile, highly integrated radio technology. Radio control points are exposed to higher software layers to make radios and applications more adaptable to changing needs and conditions. Complementary metal oxide semiconductor (CMOS) technology (an inexpensive, low-power technology) is being pushed to its limits to achieve high-speed RF circuitry coupled to high levels of integration. Several activities are combining custom signal processing for audio and video with the radio circuitry. In these efforts radios are viewed as modular building blocks that can be combined to yield systems with different cost-performance-function attributes. Some projects are investigating the architectures of software radios, in which many of the radio functions are performed by software combined with very-high-performance processing architectures.

Network Protocols and Algorithms. This effort deals with the development of robust network architectures and techniques for rapid deployment of wireless networks. Research efforts include the development of new packet-radio routing schemes such as dynamic routing protocols for ad hoc networking. The concepts being studied are not limited to end-node mobility: Other possibilities include base-station mobility and network reconfiguration as base stations are repositioned in a battlefield scenario.

End-to-End Networking. This effort addresses how best to operate across a heterogeneous mix of underlying networks, both wireless and wired. Research areas include extensions to TCP/IP that will enable mobile users to access the Internet, satellite extensions to the Internet, and overlay wireless networking that supports mobility across diverse wireless subnetworks inside buildings and in the wider area.

Mobile Applications Support. This effort deals with the development of distributed computing techniques that will enable applications to adapt to varying network connectivity and quality of service (QoS) needs. The techniques being studied include software agents (sometimes called mediators or proxies) that adapt data representations to the capabilities of bandwidth-constrained wireless links; methods of performing computations in the wireline infrastructure on behalf of power- and display-limited portable devices such as personal digital assistants (PDAs); capabilities to move code between wired and portable nodes to provide location-dependent or new functionality when the node is poorly connected; file system structures that operate whether well connected, disconnected, or poorly connected to a wired infrastructure; event-notification protocols that enable applications to learn of changes to the underlying network connectivity and QoS; and techniques for structuring applications to exploit information about their current location.

1.3.3.2 Software-Defined Radio Research

The DOD is devoting considerable attention to designing and demonstrating software-defined radios, none of which is in production as yet. The most prominent of these initiatives is the SpeakEASY program sponsored by DARPA, the Air Force Rome Laboratory, and the Army Communication Electronics Command. The key objective of SpeakEASY is to change the paradigm for military radios. In the past, radios were based on "point designs" with negligible capabilities for functional upgrades or waveform changes—capabilities that define SpeakEASY. In phase 1 of the program, analog-to-digital (A/D) converters were used to complete the radio signal path and high-speed digital signal processors (DSPs) were used for filtering and demodulation. The key technologies demonstrated in phase 1 include digital frequency conversion and wideband signal processing.

In SpeakEASY phase 2, modular radio elements (separate modules for the analog elements, A/D converter, and DSPs) will be integrated on an open-architecture bus. The key objective of phase 2 is to demonstrate a software-defined networking radio with support for legacy and future waveform evolution using a single architecture. This approach increases production volume, reduces costs, and enhances logistical support. The open-architecture design implies that competitive bids would be sought for commercial boards, modules, and software. Other goals include the use of commercial modules in the radio and the commercialization of any functions developed specifically for the radio.

The Naval Research Laboratory has an ongoing research program focusing on a software-defined radio known as the Joint C^4I Terminal (JCIT). The JCIT grew out of an Army requirement for an advanced, helicopter-based command-and-control system. The JCIT will incorporate multiple software-defined radios for combat net, intelligence communications, and military data links on a single platform.

Also under development is the advanced communications engine (ACE), which evolved from a project sponsored by DARPA. The ACE is a software-defined digital radio with capabilities for multiple simultaneous band and channel transmissions (it has six receiving and transmitting channels). The initial prototypes demonstrate "dual-use" (i.e., both military and commercial) capabilities including those of combat net radios SINCGARS and Have Quick (a UHF system designed to provide secure air-to-air and air-to-ground communications with AJ capabilities) and commercial avionics radios such as GPS, VHF air to ground, and the aircraft communications addressing and reporting system.

A very ambitious program, Millennium, was initiated to design an ultra-wideband radio. One objective was to demonstrate extremely high speed (approximately 1 billion samples per second) A/D data converters for both military and commercial communications. After the data conversion process, all tuning, filtering, demodulation, and decoding functions are performed by software (these processes and the associated technologies are discussed in Chapter 2).

1.3.3.3 *Communications Systems Research*

Several important research programs focus on complete communication systems. The DARPA Battlefield Awareness and Data Dissemination (BADD) program combines radios, ATM routers, and various communications networks and airborne relays from the Army's digital battlefield technology development effort for the deployment of high-speed data and large-file image transfer to the forward area. The Bosnia Command and Control Augmentation program, which is phase 1 of the GBS and focuses on satellite communications, grew out of BADD testing. Phase 2 of the GBS involves the incorporation of DirecTV transponders into Navy UHF satellites. Phase 3 will provide the means for stand-alone satellite transfer of high-speed data and large-file images.

1.3.3.4 *Radio Component Research*

The DOD's Extremely Lightweight Antenna program produced a compact, lightweight (under 2 pounds), and wideband (85 MHz to 2.2 GHz) antenna. The antenna incorporates a directional wideband satellite beam as well as low-gain omnidirectional radiation patterns. The DARPA Advanced Digital Receiver Technology program was initiated to demonstrate technology elements for software-defined receivers in communications, radar, and electronic warfare. Several of these functions might be merged into one digital receiver unit.

1.3.3.5 *Small Unit Operations*

The Small Unit Operations Situational Awareness System includes a significant wireless communications component. One goal of the research is to create a radio system for exchanging information among groups of up to 12 foot soldiers operating in an area of approximately 4 km^2.

1.3.3.6 *Modeling and Simulation*

The Scalable Self-Organizing Simulations (S3) Program, supported by DARPA and the National Science Foundation, uses parallel computers to simulate communications networks. This program includes projects that create models and a library of computer programs for simulating mobility, radio propagation, and teletraffic patterns in large-scale wireless networks.

1.4 COMMERCIAL TERRESTRIAL MOBILE TELEPHONE SYSTEMS AND SERVICES

Commercial wireless communications systems have exhibited remarkable growth over the past decade (see Figure 1-2). There are currently more than 50 million U.S. cellular subscribers (Hill, 1997) and more than 34 million U.S. paging subscribers (Mooney, 1997). An estimated 17 percent of the U.S. population now has cellular service, compared to 95 percent with wireline telephone service (Hill, 1997). There are also 50 million subscribers to systems based on the global system for mobile communications (GSM) standard, the European cellular technology. Worldwide, the total number of subscribers to cellular systems is projected at just under 200 million (Hill, 1997). It should be noted that these figures, as market research estimates, are fundamentally imprecise and, moreover, tend to be volatile because of the dynamic nature of the wireless industry.

Throughout the world, wireless communication systems are enabling developing countries to provide instant telephone service to new subscribers who otherwise would have to wait years for wireline access. Although wireless users are still far outnumbered by the approximately 700 million wireline telephone users worldwide, the number of new wireless subscribers is growing 15 times faster than the wireline subscriber base, and this pace is expected to accelerate in the coming years. Analysts predict that, by the year 2010, there will be equal numbers of wireless and wireline connections throughout the world.

Wireless mobile telephone systems can be divided into three genera-

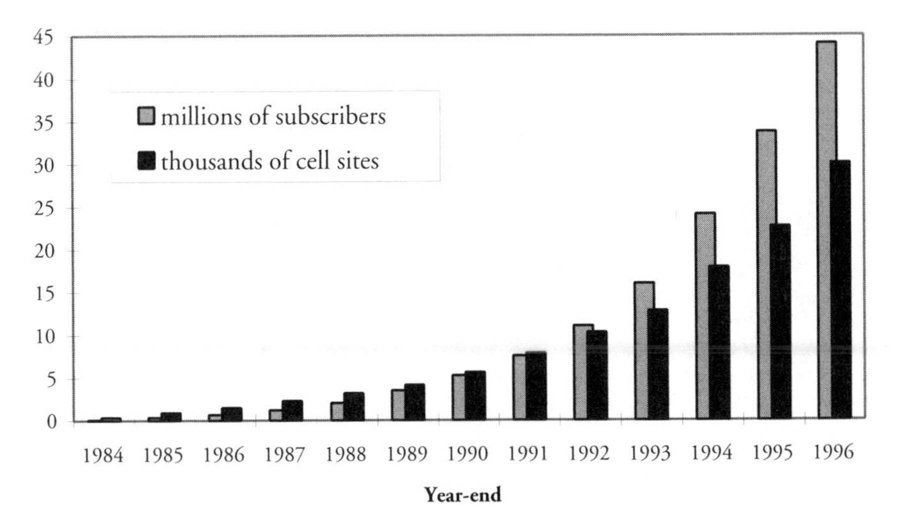

FIGURE 1-2 The number of U.S. cellular subscribers and cell sites soared between 1984 and 1996. Note that 1984 figures are for January 1985. Source: Reproduced with the Cellular Telephone Industry Association's permission from the CTIA's Semi-Annual Data Survey.

tions. The first generation, introduced in the 1980s and early 1990s, uses analog cellular and cordless telephone technology. Second-generation systems transmit speech in digital format. They provide advanced calling features and some nonvoice services. There are two categories of second-generation systems. High-tier systems feature high-power transmitters, base stations with coverage ranges on the order of kilometers, and subscribers moving at vehicular speeds. Low-tier systems, serving subscribers moving at pedestrian speeds, have low-power transmitters with a range on the order of 100 meters (m). Some of these systems are designed primarily for indoor use. Third-generation systems, planned for introduction after 2002, are expected to integrate disparate services, including broadband information services that cannot be delivered with second-generation technology. Many users are looking forward to the increased convenience promised by the integration or compatibility of systems (see Box 1-4). In addition to terrestrial mobile telephone systems, other commercial wireless systems include satellite communications, mobile data systems, and wireless local area networks (LANs).

1.4.1 First-Generation Systems

Of the original wireless communications systems deployed in the 1980s, the most popular was the analog cordless telephone, which uses

BOX 1-4
So Many Systems, So Little Integration

The proliferation of commercial communications systems can seem overwhelming, especially to international travelers. One such traveler explains: "I have a two-way pager that works in the United States. I have a one-way pager that works in some countries. I have another one-way pager that works in other countries. I've got a GSM phone. I've got a CDPD [cellular digital packet data] modem. I have a RAM [Mobile Data] and an Ardis radio. I have a cable to connect my cellular phone to the modem in my PC. I have accounts with two Internet service providers, CompuServe, America Online, an account at the office. I've got seven phone numbers in the 847 area code, one phone number in the 708 area code. I've got one phone number in New Jersey because AT&T wireless are the only people who will give you a GSM account, so I have a New Jersey phone number. I live in Chicago. . . . I have my own phone book which just has me in it. That's the problem today: I've got all of this stuff" (Lou Dellaverson, Motorola, Inc., December 10, 1996).

radio to connect a portable handset to a unit that is wired to the public switched telephone network. Hundreds of millions of such devices have been produced, and the technology has been standardized in Europe under the cordless telephone first-generation (CT0, CT1, and CT1+) standards. There is no single U.S. standard. Analog cordless telephones have ranges limited to tens of meters and require a dedicated telephone line. Cellular systems have enabled much greater mobility.

In establishing cellular service in 1983 the FCC divided the United States into 734 cellular markets (called metropolitan statistical areas and rural service areas), each with an "A-side" and "B-side" cellular service provider. Historically, the designation of A or B indicated the origins of the cellular provider: An A-side provider did *not* originate in the traditional telephone business and was called a nonwireline carrier, whereas a B-side provider had roots in traditional services and was called a wireline carrier. Each cellular carrier is licensed to use 25 MHz of radio spectrum in the 800-MHz band to provide two-way telephone and data communications for its particular market. Because the U.S. analog cellular system is standardized with AMPS, any cellular telephone is capable of working in any part of the country.

The AMPS cellular standard uses analog FM and full-duplex radio channels. The frequency division multiple access (FDMA) technique enables multiple users to share the same region of spectrum. This standard supports clear communication and inexpensive mobile telephones, but the transmissions are easy to intercept on a standard radio receiver and therefore are susceptible to eavesdropping. As of late 1996, 88 percent of all cellular telephones in the United States used the AMPS standard (digital cellular standards have only recently become available). Outside of the United States and Canada, a wide variety of incompatible analog cellular systems have been deployed (see Table 1-3). The European cellular service, which predated the AMPS system, used the Nordic mobile telephone (NMT) standard beginning in 1982. Other European nations and Japan also developed analog standards.

1.4.2 Second-Generation Systems

Spurred by growing consumer demand for wireless services, standards organizations in North America, Europe, and Japan have specified new technologies to meet consumer expectations and make efficient use of allocated spectrum bands. These second-generation systems use advanced digital signal processing, compression, coding, and network-control techniques to conserve radio bandwidth, prevent eavesdropping and unauthorized use of networks, and also support additional services (e.g., voice mail, three-way calling, and text transmission retrieval).

In the United States, second-generation technologies have been deployed in the original 800- MHz cellular bands and in personal communications bands around 1900 MHz that were allocated by the FCC between 1995 and 1997. In Europe and most other parts of the world, second-generation technologies are deployed in the 900-MHz cellular bands and in 1800-MHz personal communications bands. Japan operates digital cellular systems in various bands between 800 MHz and 1500 MHz as well as a personal communications band near 1900 MHz.

The most widespread second-generation techniques include three high-tier standards: the European standard, GSM; and two North American standards, IS-136, a time division multiple access (TDMA) technique, and IS-95, a code division multiple access (CDMA) technique.[5] The GSM standard, which has been adopted in more than 100 countries, specifies a complete wide-area communications system. The other two standards specify only the communications between mobile telephones and base stations. A separate standard, IS-41, governs communications between mobile switching centers and other infrastructure elements in the United States. Table 1-4 summarizes the properties of the principal high-tier second-generation systems.

Among low-tier standards, the personal handyphone system (PHS) provides mobile telephone services to several million Japanese subscribers. Two other standards, digital European cordless telecommunications (DECT) and cordless telephone second generation (CT2), form the basis of several wireless business telephone (i.e., private branch exchange, or PBX) products. A fourth low-tier system is the personal access communications system (PACS), a U.S. standard. Although PACS has attracted considerable industry interest, it has not been widely deployed to date. Table 1-5 summarizes the properties of low-tier systems.

TABLE 1-3 Analog Cellular Systems

Standard	Transmission Frequency (MHz) Mobile Station	Transmission Frequency (MHz) Base Station	Channel Spacing (kHz)	Regions Covered	Comments
AMPS[a]	824–849	869–894	30	America, Australia, SE Asia, Africa	
TACS[b]	890–915	935–960	25	Europe	Bands later allocated to GSM[c]
ETACS[d]	872–905	917–950	25	United Kingdom	
NMT[e] 450	453–457.5	463–467.5	25	Europe	
NMT 900	890–915	935–960	12.5	Europe, Africa, SE Asia	Frequency overlapping
C-450	450–455.74	460–465.74	10	Germany, Portugal	
RTMS[f]	450–455	460–465	25	Italy	
Radiocom 2000	192.5–199.5	200.5–207.5	12.5	France	First two bands are regional, second two are national
	215.5–233.5	207.5–215.5			
	165.2–168.4	169.8–173			
	414.8–418	424.8–428			
NTT[g]	925–940	870–885	25/6.25	Japan	First band is nationwide, others are regional
	915–918.5	860–863.5	6.25		
	922–925	867–870	6.25		
JTACS[h]	915–925	860–870	25/12.5	Japan	All are regional
NTACS[i]	898–901	843-846	25/12.5		
	918.5–922	863.5–867	12.5		

[a]Advanced mobile phone system.
[b]Total access communications system.
[c]Global system for mobile communications.
[d]Extended total access communications system.
[e]Nordic mobile telephone.
[f]Radio telephone mobile system.
[g]The dominant telecommunications operating company in Japan.
[h]Japanese total access communications.
[i]Narrowband total access communications.

SOURCE: Reprinted from Padgett et al. (1995) with permission.

TABLE 1-4 High-Tier Digital Cellular Systems

System	IS-95	GSM[a]	IS-136	PDC[b]
Region	Worldwide	Worldwide	Americas	Japan
Access method	CDMA[c]	FDMA[d]/TDMA[e]	FDMA/TDMA	TDMA
Frequency bands (megahertz)	824–849, 869–894, 1850–1910, 1930–1990	890–915, 935–960 1710–1785, 1805–1885 1850–1910, 1930–1990	824–849, 869–894	810–826, 940–956, 1477–1489, 1429–1441, 1501–1513, 1453–1465
Carrier spacing (kilohertz)	1250	200	30	25
Channels per carrier	Soft capacity (limited by noise and interference)	8	3	3

[a]Global system for mobile communications.
[b]Pacific digital cellular.
[c]Code division multiple access.
[d]Frequency division multiple access.
[e]Time division multiple access.

SOURCE: Reprinted from Padgett et al. (1995) with permission. Copyright © 1995 by IEEE.

In addition to the 1900-MHz licensed personal communications bands (see Table 1-5, the fifth column), the FCC has allocated the 1910–1930 MHz band for unlicensed low-tier systems. Commercial products based on DECT, PHS, and a modified version of PACS (designated PACS-UB, for unlicensed band) are under consideration for deployment in the 1910–1930 MHz band.

Each of the second-generation systems has distinct features and limitations, but none was designed specifically with the problems of large, complex organizations such as the military in mind. Nevertheless, it is possible to combine disparate approaches in a customized network built to meet the unique voice and data communications needs of an organization with national reach (see Box 1-5).

BOX 1-5
Tracking Packages Across North America

TotalTrack was established by the United Parcel Service (UPS) and a consortium of more than 100 cellular carriers in the United States and Canada in response to customer demands for real-time package tracking. The system was the first nationwide cellular data service. In conjunction with the private UPS telecommunications network (UPSnet), TotalTrack provides broad coverage, enabling 60,000 UPS vehicles in the United States and Canada to transmit status information to the UPS mainframe computer within minutes of package delivery. TotalTrack uses existing cellular technology and infrastructure to process 1.25 million calls and large quantities of data daily.

The UPS drivers record package information using a custom-built, handheld electronic data collection device, which is used to scan the package bar code and to capture the receiver's signature. This information is transmitted through a modem in the vehicle to the local cellular network, which provides the link to UPSnet. The system is designed to be fail-safe with cellular redundancies, dual access to UPSnet, and multiple connections to the data center.

The effectiveness of cellular technology for this application was proven by a 10-city field test that compared specialized mobile radio with cellular. Initially there were concerns about cellular reliability for data transmission. Cellular was believed to be too noisy and prone to signal interference to transmit data effectively. However, UPS achieved link reliability by using a particular combination of error-control protocols. To reduce the duration and cost of data calls, the cellular carriers connected their switching systems directly into UPSnet using a multipurpose access platform. This equipment receives the cellular data from UPS vehicles, converts it from an analog circuit-switched to a digital packet-switched format, and then forwards it to one of 40 UPS packet switches around the country. Other innovations include a phone numbering plan that allows UPS vehicles to roam between the service areas of adjacent alliance members, a billing system that consolidates all carrier charges into a single UPS bill, and a unified "help desk" that quickly resolves cellular service problems.

TABLE 1-5 Low-Tier Wireless/Personal Communications Systems

System	CT2/CT2+[a]	DECT[b]	PHS[c]	PACS[d]
Region	Europe, Canada	Europe	Japan	United States
Access method	FDMA[e]	FDMA/TDMA[f]	FDMA/TDMA	FDMA/TDMA
Frequency band (megahertz)	864–868, 944–948	1880–1900	1895–1918	1850–1910, 1930–1990
Carrier spacing (kilohertz)	100	1728	300	300
Number of carriers	40	10	77	16 per pair
Channels per carrier	1	12	4	8 per pair

[a]Cordless telephone second generation.
[b]Digital European cordless telecommunications.
[c]Personal handyphone system.
[d]Personal access communications system.
[e]Frequency division multiple access.
[f]Time division multiple access.

SOURCE: Reprinted from Padgett et al. (1995) with permission. Copyright © 1995 by IEEE.

The commercial success of second-generation wireless telephone systems has stimulated widespread interest in enhancing their capabilities to meet public expectations for advanced information services. For example, new speech-coding techniques offering improved voice quality have been introduced to all three high-tier systems. Efforts are also under way to make these systems more attractive for data services. Accordingly, standards for fax-signal transmission have been established, and standards for circuit-switched data transmission at rates of up to 64 kilobits per second (kbps) are under development for GSM and CDMA. In addition, technology for packet-switched data transmission, suitable for providing wireless Internet access, is being developed for all second-generation systems. The technology base will continue to grow as R&D organizations worldwide design innovations for a third generation of wireless communications systems.[6]

1.4.3 Third-Generation Systems

The original concept for third-generation wireless systems emerged from an International Telecommunications Union (ITU) initiative known as the future public land mobile telecommunication system (FPLMTS).[7] Over the past decade the ITU advanced the concept of a wireless system that would encompass technical capabilities a clear step above those of second-generation cellular systems. The current name for the third-generation system is International Mobile Telecommunications-2000 (IMT-2000). The number refers to an early target date for implementing the new technology and also the frequency band (around 2000 MHz) in which it would be deployed.

As envisioned in the IMT-2000 project, the third-generation wireless system would have a worldwide common radio interface and network. It would support higher data rates than do second-generation systems yet be less expensive. It would also advance other aspects of wireless communications by reducing equipment size, extending battery life, and improving ease of operation. In addition, the system would support the services required in developing as well as developed nations. Box 1-6 lists the complete set of goals established in 1990 for FPLMTS.

Since 1990 IMT-2000 recommendations have been approved that elaborate on the initial goals, establish security principles, prescribe a network architecture, present a plan for developing nations, establish radio interface requirements, and specify a framework for a satellite component. The ITU anticipated an international competition leading to a radio interface that could be developed and deployed by the year 2000. The competing radio interfaces would provide minimum outdoor data rates of 384 kpbs and an indoor rate of 2 Mbps. Other than providing a forum for discussion of

BOX 1-6
Goals for Third-Generation Commercial Wireless Systems

High quality and integrity comparable to the fixed network
Flexibility for evolution
Use of a small pocket terminal worldwide but accommodation of other terminal types
Higher service quality, especially for voice
Availability of a range of voice and other services, including multimedia
Flexible radio bearer leading to improved spectral efficiency and lower cost per erlang
Higher bit rate capability
Improved security
Improved ease of operation
Compatibility of services within the system and with the fixed network
A framework for continuing expansion of mobile network services and access to the fixed network
Integration of satellite and terrestrial components
Wider range of operating environments, including aeronautical and maritime
Open architecture that will permit easy introduction of advances in technology and applications
Services provided by more than one network in each coverage area
Services provided over a wide range of user densities and coverage areas
Services provided to both mobile and fixed users in urban, rural, and remote regions
Modular structure to enable the system to grow in size and complexity as needed
Caters to the needs of developing countries
Equipment compatible with off-the-shelf products worldwide
Service creation and service profile management by "intelligent" network
Coherent systems management
Efficient use of the radio spectrum consistent with provision of services at acceptable costs
Expanded marketplace leading to lower costs
Global standard promoting a high degree of design commonality while incorporating a variety of systems
Worldwide common frequency band
Worldwide roaming based on terminal mobility

SOURCE: International Telecommunications Union Task Group 8/1 (1996).

standards proposals, the ITU has not adopted clear plans of how to proceed beyond the point of reviewing the proposals.

The 1995 World Radio Conference set aside spectrum for nations to consider for the deployment of IMT-2000. The bands are 1920–1980 MHz and 2110–2170 MHz for terrestrial communications and 1980–2010 MHz and 2170–2200 MHz for satellites. As noted in Table 1-4 and Table 1-5, the United States has already allocated spectrum bands to personal communications that include part of the lower IMT-2000 band, making it unlikely that U.S. service providers could deploy IMT-2000 at all. Early on, attention to the ITU work was limited in both Europe and the United States, where growth in second-generation digital cellular and personal communications markets has been strong. It was the Japanese, virtually alone among all nations, who insisted that the ITU program proceed as fast as possible because they were running out of spectrum for their cellular and personal communications systems.[8] The Japanese were able to keep the IMT-2000 program on schedule, resulting in an ITU call for radio-interface proposals, now due in mid-1998. In support of this effort, the Japanese radio standards group is developing one or more Japanese standards for use in the ITU-2000 spectrum. Presumably the standard(s) will be submitted to the ITU for possible worldwide use.

Meanwhile, the European telecommunications industry established a framework for developing third-generation mobile wireless technology. The universal mobile telephone system (UMTS) is intended to replicate the commercial success achieved a decade earlier with GSM. The UMTS schedule calls for establishing the technology base by December 1997, deploying a minimum system in 2002, and achieving a full system in 2005. The technical goals of UMTS closely resemble many of the IMT-2000 goals. The Europeans plan to propose the technologies adopted for UMTS as candidates for IMT-2000.

In the United States, action on this issue did not take place until mid-1997, when the four U.S. CDMA cellular infrastructure manufacturers—Lucent Technologies, Motorola, Nortel, and QUALCOMM, Inc.—announced a third-generation program called Wideband cdmaOne. Like many candidate systems under consideration in Europe and Japan, the U.S. system uses a 5-MHz CDMA signal, although the operating parameters and design features differ from those of foreign counterparts. Additional U.S. proposals for IMT-2000 could emerge from other communities of companies supporting other digital radio interface standards.[9]

Among related developments, interest in "nomadicity" is growing within the Internet community in the United States. As originally conceived, the national information infrastructure (NII) placed little emphasis on the wireless delivery of information to mobile users (Computer Science and Telecommunications Board, 1994). But with the growth in demand for Internet services, reflected by the transition to private suppliers, providers are seeking to leverage Internet technology either directly or as part of heterogeneous networks. Plans are being made to accommodate nomads (i.e., mobile users) who draw on a variety of communications, computing, and information systems simultaneously, a concept that will require attention by multiple industries to issues such as security, interoperability, and synchronization within and between systems (Cross-Industry Working Team, 1995).

Other ITU activities are addressing network aspects of IMT-2000.[10] Here again the Japanese have made major contributions toward the establishment of a single worldwide network to support wireless systems. Only in mid-1997 did the U.S. and European delegations begin to make significant contributions, concerned about their current investments in cellular and personal communications networks and the possible effects of establishing a worldwide network that was incompatible with their systems. The latest U.S. and European proposals emphasize the idea of a family of networks supporting a family of radio interfaces through the use of appropriate gateways to achieve worldwide roaming and interoperability.

Although it is clear that many new wireless communications technologies will emerge in the 2002–2005 time frame, it is not clear when and how they will be commercialized. The robust evolution of second-generation systems will limit commercial incentives to introduce a new generation of systems. It is possible that advances in second-generation systems will meet future demand for mobile telephone services and that a demonstrated demand for high-bit-rate data services will be necessary to stimulate the commercial deployment of third-generation technology.

1.5 COMMERCIAL SATELLITE SYSTEMS

Satellite systems can be classified by frequency and orbit. Above 1 GHz a satellite signal easily penetrates the ionosphere. Transmission at higher frequencies is desirable because additional bandwidth is available there, but then expensive components are needed to overcome signal attenuation, absorption, and path loss (see Chapter 2 for a discussion of channel impairments). Most satellite systems are of the GEO variety, offering configuration simplicity, wide footprint (i.e., one satellite covers an entire geographical region), and fixed satellite-to-ground-terminal characteristics. But GEO systems also have a number of disadvantages, including long propagation delays (a round-trip takes approximately half a second), high transmitter-power requirements, and poor coverage at the far northern and southern latitudes. Moreover, GEO satellites are expensive to launch, and, because only a handful of satellites are typically used to achieve global coverage, they are vulnerable to single points of failure.

The International Maritime Satellite (INMARSAT) Organization, formed in 1979, is now backed by the governments of 75 member countries. Its first satellites (INMARSAT-A) became operational in 1982, supporting voice and low-rate data applications with analog FM technology. By the end of 1993, 30,000 ground terminals were in operation. The next generation of INMARSAT satellites (INMARSAT-B and C) used digital technology, but data rates remained low (600 bps). With the introduction of INMARSAT-M in 1996 it is now possible to use laptop computer-sized satellite terminals for voice and low-rate (2.4 kbps) data transmission. However, the voice quality of this system remains poor due to propagation delay, and data transmission rates are 10 times slower than those of a standard modem.

In the late 1980s QUALCOMM deployed the OMNITracs vehicle-tracking and communications system for both North America (using GSTAR satellites) and Europe (using EUTELSAT satellites). The service provides two-way messaging and automatic position reporting. By 1997 more than 200,000 trucks, most of them in the United States, were equipped with the system. The use of such systems in Europe has been restricted by high equipment costs and expectations for less-costly alternatives with the next generation of systems.

Recently introduced GEO systems for data communications include Mobilesat in Australia and MSAT in North America (see Table 1-6). Innovations in GEO systems include spot beams for custom broadcast coverage and improved on-board processing. Although GEO satellite communications systems are not fully mobile (i.e., the terminals are not handheld), innovations in terminal design have enabled the development of private networks and rapidly reconfigurable systems. Very small aperture terminals (VSATs) use small Earth-station antennas to form private networks through links to GEO satellites. The VSAT is the result of more than 20 years of advances in digital Earth-station technology. The applications have evolved from point-to-point transmission links to networking terminals that leverage the broadcasting capability of satellites.

TABLE 1-6 Selected Geosynchronous Earth Orbiting Systems

System	Organization	Number of Satellites	Coverage	Data Rate (kbps)[a]
MSAT[b]	AMSC[c], TMI[d]	2	North America	4.8
INMARSAT-M	INMARSAT	5	Global	2.4
Mobilesat	Optus Communication	2	Australia	2.4
EMS[e]	European Space Agency	1	Europe, Northern Africa	10
LLM[f]	European Space Agency	1	Europe, Asia	10

[a]Kilobits per second.
[b]Mobile satellite.
[c]American Mobile Satellite Corporation.
[d]Telesat Mobile, Inc.
[e]European mobile satellite.
[f]L-band land mobile.

SOURCE: Reprinted from Abrishamkar and Siveski (1996) with permission. Copyright © 1996 by IEEE.

The VSAT terminals offer various types of access. Fast-response protocols are used for time-sensitive transactions such as credit card purchases and hotel or airline reservations, throughput-efficient access is used for file transfers, and circuit-switched access is used for speech and digital video. (Throughput is the fraction of time during which a channel can be used.) An important feature of VSAT technology is ease of deployment: Installation takes approximately 2 hours. Companies are now installing VSATs at the rate of more than 1,500 per month. There are more than 200,000 VSATs worldwide, operating in nearly every country; individual networks range in size from as few as 20 nodes operating in a shared-hub environment to nearly 10,000 in the General Motors Corporation network.

In 1994 direct-broadcast satellites (DBSs) became operational, some two decades after the first experiments were performed with this technology. These systems broadcast a signal from a GEO satellite with sufficient power to allow direct reception in a home, office, or vehicle with an inexpensive receiver. The two primary applications for DBS systems are television and radio; emerging applications include DirecPC and GBS. Systems for direct-broadcast television are operational in Europe, Japan, and the United States. By the end of 1996 these systems had more than 2.5 million U.S. subscribers. Digital audio broadcasting (DAB) has the potential to provide every radio within a service area with continuous transmissions of a sound quality comparable to that of a compact disc. Systems are being tested around the world that deliver DAB from satellites as well as from terrestrial antennas.

Communications systems using non-GEO satellites are emerging as major players in commercial wireless applications. These satellites are characterized as either medium Earth orbit (MEO) or low Earth orbit (LEO). The LEOs, deployed in either circular or elliptical orbits of 500 to 2,000 km, offer several advantages including reduced propagation delay and low transmit-power requirements, allowing the use of handheld terminals. But at these altitudes a system requires many satellites to achieve global coverage. Furthermore, satellite movement relative to the ground terminal introduces Doppler shift in the received signal, and each satellite is visible from a ground terminal for only a few minutes at a time so that handoffs between satellites are frequent. The MEO satellites offer features that represent a compromise between LEOs and GEOs. The MEOs are deployed in circular orbits at an altitude of about 10,000 km. Approximately 10 to 15 satellites (more than GEOs but fewer than LEOs) are required for global coverage, and average visibility is one to two hours per satellite (less than for GEOs but more than for LEOs). The Doppler shift in MEOs is also considerably less than that in LEOs, but higher transmit power is required.

The majority of new satellite systems that will become operational by the year 2000 are LEO or MEO systems. These satellites can be categorized further by size. Big LEO/MEOs (see Table 1-7) support voice and data communications with large satellites (weighing 400–2,000 kilograms [kg]) and operate at frequencies above 1 GHz. Little LEOs use much smaller satellites (weighing 40–100 kg) and operate in the UHF and VHF bands, thereby enabling the use of inexpensive transmission hardware for both the satellite and ground terminal. The 36-satellite Orbcomm system is an example.

Most of these systems provide voice and low-rate data to mobile users with handheld terminals. The link rates for little LEOs are asymmetric, with lower rates on the uplink (ground to satellite) than on the downlink (satellite to ground) because of power limitations in the handheld unit. Teledesic is unusual because it is intended primarily for broadband wireless data communications with stationary terminals at integrated services digital network (ISDN) rates. Teledesic and Iridium have direct intersatellite communication links independent of the ground segment, enabling the provision of services to countries lacking a communications infrastructure. Iridium is designed to consist of 66 satellites arranged in six planes, all in a nearly polar orbit. Each satellite is expected to serve as a "switchboard in the sky," routing each channel of voice traffic through various other satellites in the system; communications are eventually delivered to an appropriate ground-based gateway to terrestrial telecommunications.

Globalstar is a LEO digital telecommunications system that will begin offering wireless telephone, data, paging, fax, and position location services worldwide beginning in 1998. The 48-satellite constellation operating 1,410 km from the planet surface serves as a "bent-pipe" relay to local ground-based infrastructure.

1.6 MOBILE DATA SERVICES

Commercial packet-switched mobile data services emerged after the success of short-message, alphanumeric one-way paging systems. Mobile data networks provide two-way, low-speed, packet-switched data communication links with some restrictions on the size of the message (10 to 20 kilobytes) in early systems. Services provided by mobile data networks include the following:

- Transaction processing (credit card verification, paging, notice of voice or electronic mail);
- Broadcast services (general information, weather and traffic advisories, advertising);

TABLE 1-7 Selected Big Low Earth Orbit (LEO)/Medium Earth Orbit (MEO) Systems

System	Organization	Number of Satellites	Orbit	Coverage	Data Rate (kbps)[a]	Year Operational
Globalstar	Loral/QUALCOMM	48	LEO	Global	9.6	1998
Iridium	Motorola	66	LEO	Global	2.4	1998
Odyssey	TRW	12	MEO	Global	9.6	1998
Teledesic	Teledesic	240	LEO	Global	20–2,000	2002
ICO	ICO Global Communications	10	MEO	Global	2.4	1999
Archimedes	European Space Agency	5–6	MEO	Europe, Asia, Canada	256	After 2000

[a] Kilobits per second.

SOURCE: Reprinted from Abrishamkar and Siveski (1996) with permission. Copyright © 1996 by IEEE.

- Interactive services (terminal access to host, remote LAN access, games); and
- Multicast service (subscription information services, law enforcement, private bulletin boards).

The first commercial mobile data network was Ardis, a private network developed in 1983 by IBM Corporation and Motorola to enable IBM to provide computing facilities in the field. By 1990 Ardis was deployed in more than 400 metropolitan areas and 10,700 cities and towns using 1,300 base stations. By 1994 Ardis (since then owned by Motorola) provided nationwide roaming for approximately 35,000 users, at a rate of 45 million messages per month, and a data rate of 19.2 kbps.

In 1986, Swedish Telecomm and Ericsson Radio Systems AB introduced Mobitex and deployed it in Sweden. This system is available in the United States, Norway, Finland, Great Britain, the Netherlands, and France. The system supports a data rate of 8 Mbps and nationwide roaming (international roaming is planned). This service is distributed by RAM Mobile Data in the United States, where by 1994 it had 12,000 subscribers. A total of 840 base stations are connected to 40 switching centers to cover 100 metropolitan areas and 6,300 cities and towns.

Cellular digital packet data (CDPD) technology was developed by IBM, which together with nine operating companies formed the CDPD Forum to develop an open standard and multivendor environment for a packet-switched network using the physical infrastructure and frequency bands of the AMPS systems. The CDPD specification was completed in 1993 with key contributions from IBM, McCaw Cellular Communications, Inc., and Pacific Communications Sciences, Inc. Deployment of the 19.2-kbps CDPD infrastructure, designed to make use of idle channels in analog cellular systems, commenced in 1995.

In the 1990s Metricom, Inc., developed a metropolitan-area network that was deployed first in the San Francisco Bay area and then in Washington, D.C. The signaling rate of this system is advertised at 100 kbps but the actual data rate is substantially slower. The Metricom system uses "frequency hopping" spread-spectrum (FHSS) technology in the lower frequencies (around 900 MHz) of the unlicensed industrial, scientific, and medical (ISM) bands.[11]

In 1996 the European Telecommunications Standards Institute (ETSI) standard for mobile data services, trans-European trunked radio (TETRA), was completed. It is currently being used primarily for public safety purposes. Work is in progress to enhance the digital cellular and personal communications technologies. More recently, the digital cellular standards (GSM, IS-95, PHS, PACS, and IS-136) have been updated to support packet-switched mobile data services at a variety of data rates. Key features of existing mobile data services are shown in Table 1-8. Although many services are available, the mobile data market has grown more slowly than have voice services.

1.7 WIRELESS LOCAL AREA NETWORKS

Wireless LANs provide data rates exceeding 1 Mbps in coverage areas with dimensions on the order of tens of meters. They are used for a variety of applications, including the following:

- LAN extensions in hospitals, factory floors, branch offices, and offices with wiring difficulties;
- Cross-building inter-LAN bridges that serve as point-to-point, high-speed links connecting separate LANs located within a few miles of each other;
- Temporary ad hoc networks used in conference registration, campaign headquarters, and military camps;
- Temporary wireless access to a wired LAN from a portable device such as a laptop computer; and
- Access to centralized computing facilities of a shipboard or research facility through a wireless device such as a notepad computer.

In 1990 the Institute of Electrical and Electronics Engineers (IEEE) formed a committee to develop a standard for wireless LANs operating at 1 and 2 Mbps. In 1992 the ETSI chartered a committee to develop a standard for high-performance radio LANs (HIPERLAN) operating at 20 Mbps.

Table 1-9 indicates the technical features of various LAN products (including some that use the infrared portion of the spectrum and are therefore not examined in detail in this report). The market for wireless LAN products is growing rapidly but not nearly as fast as the market for wireless voice applications. The $200 million market for wireless LANs is tiny compared to the cellular industry, which is worth billions (Wickelgren, 1996).

1.8 COMPARISON OF INTERNATIONAL RESEARCH, DEVELOPMENT, AND DEPLOYMENT STRATEGIES

Commercial wireless technologies have followed divergent evolutionary paths in different parts of the world. For example, strong contrasts are evident in the transition from first-generation cellular systems to second-generation systems in the United States and Europe. At first a single U.S. system was used for analog cellular communications, AMPS, and every cellular telephone in the United States and Canada could communi-

TABLE 1-8 Mobile Data Services

System	Ardis	Mobitex	CDPD[a]	TETRA[b]	Metricom
Frequency band (MHz)[c]	800 bands; 45 kHz[d] sep.	935–940, 896–961	869–894, 824–849	380–383, 390–393	902–928 (ISM[e] bands)
Channel bit rate (kbps)[f]	19.2	8.0	19.2	36	100
RF[g] channel spacing (kHz)	25	12.5	30	25	160
Channel access/ multiuser access	FDMA[h]/ALOHA	Slotted ALOHA	FDMA/ALOHA	ALOHA	FHSS[i]/BTMA[j]

[a]Cellular digital packet data.
[b]Trans-European trunked radio.
[c]Megahertz.
[d]Kilohertz.
[e]Industrial, scientific, and medical.
[f]Kilobits per second.
[g]Radio frequency.
[h]Frequency division multiple access.
[i]Frequency hopping spread spectrum.
[j]Busy tone multiple access.

SOURCE: Reprinted from Cox (1995) with permission. Copyright © 1995 by IEEE.

TABLE 1-9 Wireless Local Area Network Technologies

Technology	DFIR[a]	DBIR[b]	RF[c]	DSSS[d]	FHSS[e]
Channel bit rate (Mbps)[f]	1–4	10–155	5–10	2–20	1–3
Mobility	Stationary/ portable	Stationary with LOS[g]	Stationary	Stationary/portable	Portable
Range (meters)	15–60	30	10–40	30–200	30–100
Frequency bands	Infrared	Infrared	18 GHz[h], ISM[i]	ISM	ISM
Systems (companies)	Spectrixlite (Spectrix Corp.); Photolink (Photonics)	Infralan (InfraLAN); UWIN (Jolt Ltd.)	Altair (Motorola, Inc.); Fast Wave (Southwest Microwave, Inc.); RediCARDrf (Data Race, Inc.)	Roamabout (Digital Equipment Corp.); ARLAN (AiroNet Wireless Communications); WaveLAN (Lucent Technologies); INTERSECT (Persoft, Inc.); AIRLAN (Solectek Corp.); RangeLAN (Proxim); FreePort (WinData); PRISM (Harris Corp.)	Range-LAN2 (Proxim); PortLAN (RDC Networks); Netwave (Xircom)

[a]Diffused infrared.
[b]Directed-beam infrared.
[c]Radio frequency.
[d]Direct sequence spread spectrum.
[e]Frequency hopping spread spectrum.
[f]Megabits per second.
[g]Line of sight.
[h]Gigahertz.
[i]Industrial, scientific, and medical.

SOURCES: Reproduced from material in Cox (1995) and Pahlavan et al. (1995).

cate with every base station. By contrast, European users were faced with a complex mixture of incompatible analog systems. To maintain mobile telephone service, an international traveler in Europe needed up to five different telephones. The situation was reversed by second-generation systems. Now there is a single digital technology, GSM, deployed throughout Europe (and in more than 100 countries worldwide), whereas the United States has become a technology battleground for three competitors: GSM (DSC-1900), TDMA (IS-136), and CDMA (IS-95).

The differences in technology evolution are due in large measure to different government policies in Europe, the United States, and Japan, the world's principal sources of wireless technologies. Three types of government policies influence developments in wireless systems: policies on radio spectrum regulation, approaches to R&D, and telecommunications industry structure. The reasons for the shifts in the above example can be found primarily in changes in spectrum regulation policies adopted in the 1980s. In establishing first-generation systems in the United States in the late 1970s, the FCC regulated four properties of a radio system: noninterference, quality, efficiency, and interoperability. In the 1980s, deregulation was in vogue and the scope of the FCC's authority was restricted to noninterference; the other properties were deemed commercial issues to be settled in the marketplace. Although this policy stimulated innovation in the U.S. manufacturing industry, it also meant that operating companies had to choose among various competing technologies.

In Europe, the main trend in government regulation in the 1980s was a move from national authority to multinational regulation under the aegis of the European Community (EC; now the European Union [EU]). The EC had a strong interest in establishing continental standards for common products and services, including electric plugs and telephone dialing conventions. In this context the notion of a telephone that could be used throughout Europe had a strong appeal. To advance this notion, the EC offered new spectrum for cellular service on the condition that the operating industries of participating countries agree on a single standard. Attracted by the availability of free spectrum, operating companies (many of them government-owned) in 15 countries put aside national rivalries and adopted the GSM standard.

Thus, a new pattern of technical cooperation was established in Europe. This cooperation was reinforced by the European Commission (the administrative unit of the EU), which funded cooperative precompetitive research focusing on advanced communications systems, first in the Research for Advanced Communications in Europe (RACE) program and then in the Advanced Communications Technologies and Services (ACTS) program. In both programs a consortium of companies and universities performs the research. Spectrum management rules continue to prescribe a single standard for each service, meaning that an industry consensus is required before a standard is introduced. Once a technology is established, companies enter the competitive phase of product development and marketing. This process promotes a thorough investigation of technologies prior to standardization and assures economies of scale when commercial service begins. In preparation for UMTS, scheduled for initial deployment in 2002, extensive R&D and evaluation of competing prototypes have been under way since 1994. All of this activity will provide European industry with a strong technical base for realizing the goals for mobile communications in the first decade of the next century.

The U.S. approach to communications technology R&D is much more competitive. Individual companies perform much of this research in the context of their product marketing plans. Coordination takes place within diverse standards organizations such as the Telecommunications Industry Association, IEEE, and American National Standards Institute. Some interaction also takes place in the GloMo program, which brings together universities and industry to fill specific technology gaps identified by DARPA program managers. But for the most part standards setting is a competitive rather than cooperative process, with each company or group of companies striving to protect commercial interests. The FCC rules for spectrum management allow license holders to transmit any signals, subject only to constraints on interference with the signals of other license holders. Similar flexibility is extended to unlicensed transmissions. As a consequence, there are multiple competing standards (seven in the case of wideband personal communications) for wireless service in the United States.

Government policies on industry structure also strongly influence technology development. After the FCC issued cellular operating licenses, most of the companies that began offering cellular service had limited technical resources and relied almost entirely on vendors and consultants for technical expertise. Even the cellular subsidiaries of the regional Bell operating companies had to build a new base of expertise: Under the terms of the consent decree that broke up AT&T in 1984, these cellular companies had no access to the abundant technical resources of Bellcore, the research unit of the regional Bell companies. In this environment, much of the new wireless communications technology in the United States has come from the manufacturing industry, with the result that proprietary rather than open network-interface standards have proliferated. The published technical standards for wireless communications were at first confined to the air interface between terminals and base stations. Eventually the industry adopted a standard for intersystem operation to facilitate roaming. Many other interfaces, especially those between switching centers and base stations, remain proprietary but the situation is changing to allow fully open systems.

By contrast, the European cellular operating industry has been dominated by national telephone monopolies. These companies have strong research laboratories that participate fully in technology creation and standards setting. To gain the advantage of flexibility in equipment procurement, operating companies favor mandatory open interfaces, a preference reflected in the GSM standard.

Little has been published concerning the factors that influence the evolution of wireless communications technology in Japan. In recent years NTT, the dominant telecommunications operating company, has provided a strong coordinating mechanism for creating and standardizing new technology. The biggest success has been PHS, which entered commercial service in 1995 and attracted 4 million subscribers in its first year of operation. The initial R&D for PHS was conducted by NTT, but it licenses many manufacturers to offer PHS equipment. Now many Japanese companies are cooperating in a study of wideband CDMA technology for third-generation systems. A joint experimental trial of one system is scheduled for the end of 1997. In addition to corporate R&D, a government organization, Research and Development Center for Radio Systems, is a significant source of wireless communications technology in Japan.

Worldwide efforts to guide the evolution of wireless communications technology come together in the IMT-2000 project. National delegations to IMT-2000 reflect their country's policies: The U.S. delegation pushes for diversity,[12] the Europeans advocate a structure favorable to UMTS and its descendants, and the Japanese delegation favors convergence to a small number of worldwide standards. Other countries assert their own service needs, which in some cases can be met by mobile communications satellites and in other cases by wireless local loops.

1.9 SUMMARY AND REPORT ORGANIZATION

The history of wireless communications suggests a number of key points to be considered in evaluating potential future strategies for the DOD and DARPA. Wireless technology has now evolved to a point where the goal of "anytime, anywhere" communications is within reach. Since 1980 consumer demand for cordless and cellular telephones has driven rapid growth in wireless services, especially for voice communications. Wireless data services have not taken off as yet although expectations are high, given the growth of Internet applications. Extensive research is under way to develop third-generation commercial wireless systems, which are expected to be in place before 2010. These trends suggest that the DOD will continue to have an ample selection of advanced commercial wireless technologies from which to choose.

The DOD, which currently uses a variety of wireless systems based on 1970s and 1980s technology, is relying increasingly on commercial wireless products to cope with reductions in defense budgets and the growing need for flexible systems that can be deployed rapidly. In the Gulf War, the DOD used commercial equipment such as GPS receivers and INMARSAT links and found that performance was comparable to that of technologies designed explicitly to meet military needs. However, the DOD will continue to have unique needs for security, interoperability, and other features that might not be met by commercial products. The gaps between commercial technologies and military needs are difficult to identify precisely because, although the DOD has defined its vision for future untethered systems in general terms, projected operational needs have apparently not been translated into technical specifications that conform to the capabilities of commercial products.

The GloMo program and other military R&D efforts are attempting to meet DOD's future communications needs and have produced some useful results. However, none of these programs has adopted a systems approach to the problem, most notably with respect to the design of a network architecture. There may be other unmet needs as well; however, the committee based its work on first principles rather than an assessment of GloMo. A new strategy may be needed to identify the needs more specifically as a basis for determining where to focus DARPA's R&D efforts and where commercial products will suffice.

The effort to evaluate commercial technologies in light of defense needs will be complicated by the characteristics of the U.S. marketplace. In Europe there is a single standard (GSM) for digital wireless communications, and precompetitive research on new wireless technologies is carried out in cooperative, government-funded programs. The U.S. wireless market features a mixture of competing standards, and most technology R&D is conducted by individual companies. This environment forces operators to choose from an assortment of competing technologies.

The remainder of this report is an attempt to help the DOD devise strategies for making those choices. Chapter 2 provides technical background on the many issues that need to be addressed in designing wireless communications systems, which are extremely complex. The highly technical discussion may not interest all readers but is fundamental to any informed analysis of wireless systems. Chapter 3 explores the opportunities for and barriers to synergy between the military and commercial sectors in the development of wireless technologies. Chapter 4 integrates all the information presented in this report to provide a set of recommendations for the DOD and DARPA.

NOTES

1. This report does not address unguided optical communications systems, which use the 10^3–10^7 gigahertz frequency band (infrared, visible, and ultraviolet light), because the commercial products that operate in these bands are designed for indoor applications and therefore would not be of great use in military applications.

2. A protocol is a set of rules, encoded in software, for performing specific functions.

3. The developments since the mid-1970s, when the use of computer networks moved beyond the ARPA research community, paved the way for commercial services. The CSNet project, funded by the National Science Foundation (NSF) for the computer science community, eventually led to the NSFNET and a dramatic increase in the number of interconnected nodes. The commercialization of Internet service was symbolized by the decommissioning of the ARPANET in 1990 and privatization of the NSFNET in 1995.

4. Two types of codes are used to spread the signal. A long code is reserved for use by the military to obtain location information within a few meters of accuracy and timing information within 100 nanoseconds. A shorter code is used by commercial systems to obtain location information accurate to within 100 meters.

5. A fourth digital modulation technique, based on Motorola's iDEN technology, is used by some specialized U.S. mobile radio services in the lower 800-MHz band to provide cellular-like voice, trunked radio, paging, and messaging services.

6. One integrated solution not addressed in detail in this report is the new generation of public safety radio networks. These systems are used in both the military and commercial sectors for applications such as law enforcement and fire fighting. Until recently these systems were characterized simply as 25-kilohertz FM voice radios and 9.6-kbps modems. In the past a municipal law enforcement radio system typically was deployed as a redundant overlay of towers and repeaters separate from the radio systems operated by fire, health, highway, and other municipal departments. Today's tight budgets often force municipalities to pool departmental funds to upgrade public safety radios and establish a single system with enough capacity to meet every user's needs. To assist in this process the Association of Public Safety Communication Officers (APCO), which includes law enforcement, highway, forestry, health, and many other municipal and federal users, recently initiated an ambitious program called Project-25 to reduce the cost of next-generation radios. APCO Project-25 seeks to reduce user dependence on proprietary radios from a single manufacturer (generally the system installer) and introduce cost competition in the upgrading and replacement market at the municipality level. The strategy is to standardize a digital-modulation radio, which would be described as APCO Project-25 compliant, thus opening up public radio purchasing to a variety of competing manufacturers. Some radios that are APCO Project-25 compliant are now available and are being adopted by the Federal Law Enforcement Radio Users Group (representing radio users in the Federal Bureau of Investigation, Drug Enforcement Agency, Secret

Service, Department of the Treasury, and other civilian agencies). The APCO Project-25 process has encouraged an unprecedented level of cooperation among municipal radio users.

7. These activities are carried out by the ITU Radiocommunication Sector (ITU-R) Working Party 8/13, later renamed ITU-R Task Group 8/1.

8. The implementation of standards based on IMT-2000 in Japan clearly would give Japanese companies early experience with the technology and perhaps position them to dominate future world markets for IMT-2000 products.

9. Although optical communications systems are not addressed in detail in this report, in large part because the commercial research focuses on indoor applications, the advantages of laser systems need to be mentioned. A laser produces optical radiation by stimulating emissions from an electronic or chemical material. Unlike light produced by incandescent or fluorescent sources, the resultant beam is coherent and exhibits extremely low angular divergence, properties that enable transmissions spanning great distances (i.e., thousands of miles). The data, voice, images, or other signals are modulated on a beam of light, which is detected by an optical receiver and decoded. The transmitter and receiver need to be in direct visual contact, and so the laser beam is steered in the appropriate direction using mirrors or other optical elements. Laser communications systems offer several advantages over RF systems. The main advantage is high capacity: Systems now under development will support transmissions in the range of hundreds of megabits per second, with systems under consideration attaining the gigabits-per-second range. Another advantage is the low power requirement for point-to-point communications (orders of magnitude lower than RF systems). All the energy is focused into a very narrow beam because the physical dispersion of a laser beam in space is minimal. Furthermore, laser communications systems offer security benefits because almost no energy is diffused outside the laser beam, which is therefore not easily detected by an adversary. This combination of features makes laser communications systems attractive for secure transmissions between hub points in mobile, dynamically changing environments (e.g., between base stations on vehicle-mounted switching facilities). However, laser systems are sensitive to interference from other light sources, such as the sun, and any obstructions of the visual link by dust, rain, or fog. There is also a risk of damage to the eyes of unprotected observers. Finally, components for laser-based systems are much more expensive than those for RF systems and therefore are unlikely to penetrate the commercial market for some time.

10. These activities are carried out by the ITU Telecommunications Sector, Study Group 11.

11. The ISM bands (at 902–928 MHz, 2400–2483 MHz, and 5700–5850 MHz) are available for any wireless device that uses less than 1 watt of transmit power.

12. The United States participates in the IMT-2000 process in Task Group 8/1 through a delegation led by the FCC.

2

Technology Limits, Trade-offs, and Challenges

Wireless communications networks incorporate a broad range of technologies, including electrochemical materials, electronic devices and circuits, antennas, digital signal processing algorithms, network control protocols, and cryptography. Although all of these technologies are well advanced in other applications, wireless systems introduce a set of constraints and challenges beyond those addressed in the evolution of other communications networks, such as the (wireline) public switched telephone network and the Internet. These special constraints make it exceedingly difficult to design affordable wireless systems that meet every need. The challenges can be grouped into three categories: mobility, connectivity, and energy.

Mobility is a fundamental feature of untethered communications networks. Portable, wireless communications devices significantly enhance the mobility of users, but they also pose network design difficulties. As the communications devices move, the network has to rearrange itself. To deliver information to a mobile terminal, the network has to learn the new location(s) of the terminal and change the routing of information accordingly, sometimes at very high speeds. The rerouting must be done seamlessly without any perceived interruption of service.

A wide variety of problems arise when mobile wireless communications terminals send and receive signals over the air. The signals of all the terminals are subject to mutual interference. The characteristics of the propagation medium change randomly as users move, and the mobile radio channel introduces random variation in the received signal power and other distortions, such as frequency shifts and the spreading of signals over time. Signals that travel over the air are also more vulnerable to jamming and interception than are those transmitted through wires or fibers. These limitations are often addressed with a combination of sophisticated signal processing techniques and antennas. However, these solutions add to the complexity of portable communications devices and increase power requirements.

Wireless systems pose two types of power challenges. First, when power is radiated from an antenna, very little of it typically reaches the receiver, a phenomenon known as path loss. This problem can be partly overcome with increased transmit power, special types of antennas, and other solutions. Second, wireless terminals often carry their own power supplies in the form of batteries. Battery life is limited and is influenced by many aspects of terminal design as well as the technology of the network infrastructure. Scarce power constrains the signal processing capabilities and transmit power of the mobile terminal, motivating efforts to keep these units as simple as possible. However, a low-power design cannot accommodate the most sophisticated techniques available to cope with the vagaries of the wireless channel and support the network protocols of mobility management. In the absence of research breakthroughs that simplify these techniques, the only solution is to increase the complexity of the network, which needs to compensate for the simplicity of portable communications devices.

The challenges related to mobility, connectivity, and energy have stimulated a high level of R&D activity in the telecommunications industry and academia. Still, a chasm remains between the capabilities of wired and wireless communications systems. Even as commercial wireless systems evolve, additional features will be needed to meet military requirements for untethered communications. Military applications introduce additional challenges because the systems need to be rapidly deployable on mobile platforms in any one of a diverse range of operating environments; they need to interoperate with other systems; and they need protection against enemy attempts to jam, intercept, and alter information.

This chapter provides the technical basis for the analysis of military-commercial synergy in Chapter 3 by examining the challenges of mobility, connectivity, and energy and the technologies devised to overcome them. The discussion refers to the various layers of a network as defined in the Open Systems Interconnection (OSI) model (see Box 2-1). Section 2.1 is a tutorial on the wireless channel, its capacity limits, techniques for overcoming channel impairments, and the access and operational issues that arise when multiple users share the same channel. The next three sections address network, system, and hardware issues with an emphasis

BOX 2-1
Open Systems Interconnection Model

The Open Systems Interconnection model identifies seven layers, some or all of which are implemented by virtually any network system. The *physical* layer includes the mechanical, electrical, and procedural interfaces to the transmission medium. The *link* layer converts the transmission medium into a stream that appears to be free of undetected errors. This layer includes error-correction mechanisms and the protocols used to gain access to shared channels. The *network* layer chooses a route from the sender to the receiver and deals with congestion and address issues. The IP protocol falls into this layer. The *transport* layer is responsible for the end-to-end delivery of data. The TCP protocol falls into this layer. The *session* layer allows multiple transport-layer connections to be managed as a single unit. The *presentation* layer chooses common representations (typically application dependent) for the data being carried. The *applications* layer deals with application-specific protocol issues.

on military needs. Section 2.2 examines network design issues including architecture, resource allocation and discovery, inoperability, mobility management, and simulation and modeling tools. Section 2.3 addresses end-to-end systems design issues including application-level adaptation, quality of service, and security. Section 2.4 reviews hardware issues of particular military concern, focusing on radio components.

2.1 COMMUNICATION LINK DESIGN

The ideal wireless communications system would provide high data rates with high reliability and yet use minimum bandwidth and power. It would perform well in wireless propagation environments despite multiple channel impairments such as signal fading and interference. The ideal system would accommodate hardware constraints such as imperfect timing and nonlinear amplifiers. The mobile units would have low power requirements and yet still provide adequate transmit power and signal processing. In addition, despite the system complexity required to achieve this performance level, both the transmitter and receiver would be affordable.

Such a system has yet to be built. In fact, many of the desired properties are mutually exclusive, meaning that trade-offs need to be made in system design. A case in point is the choice of approaches for overcoming the limitations and impairments of the wireless channel. The impairments inherent in any wireless channel include the rate at which received signal power decreases relative to transmitter-receiver distance (path loss), attenuation caused by objects blocking the signal transmission (shadow fading), and rapid variations in received signal power (flat fading). The impairments determine the types of applications that can be supported in different propagation environments. Applications require different data rates and bit-error rates (BER, or the probability that a bit is received in error). For example, voice applications require data rates on the order of 8 to 32 kbps and a maximum tolerable BER of 10^{-3}, whereas database access and remote file transfer require data rates up to 1 Mbps and a maximum tolerable BER of 10^{-7}.[1] Both sets of specifications are difficult to achieve in many radio environments.

In general, systems designed for the worst-case propagation conditions assume high error rates, which limit their capability to support high-speed data and video teleconferencing applications. The random nature of the radio channel makes it difficult to guarantee quality and performance for demanding applications. However, a wireless system can be designed to adapt to the varying link quality at both the link and network level, such that the system can support improved data rates and quality. Applications can also be designed to adapt to deteriorating channel conditions to minimize the degradation perceived by the user. The overall system can be optimized by making trade-offs among various performance measures such as BER, outage probability, and spectral and power efficiency. These trade-offs dictate the choice of modulation, signal processing, and antenna techniques used to mitigate channel impairments.

These techniques require fairly intensive digital signal processing at the mobile unit. The extent of the computation that can be performed is limited by the power available to drive the DSP chips and the microprocessor. Thus, in addition to being power limited, the mobile unit is also complexity limited, which means that trade-offs need to be made in designing the communication link. For example, the transmit power requirements of the mobile unit can be reduced if error-correction coding is used, but then additional power is needed to drive the encoding and decoding hardware. In cellular systems it is preferable to place much of the computational burden at the base station, which has fewer power restrictions than do the mobile units. Research aimed at simplifying DSP and antenna processing techniques (Section 2.4) can also help mitigate the computational burden.

The remainder of this section outlines the characteristics of the wireless channel, focusing on fading and interference problems (Section 2.1.1); key communications technologies, including modulation and coding (Sections 2.1.2 through 2.1.4); the countermeasures available to address fading and interference (Section 2.1.5); and the various ways in which users access wireless systems (Section 2.1.6).

2.1.1 Characteristics of the Wireless Channel

The characteristics of the radio channel impose fundamental limits on the range, data rate, and quality of wireless communications. The performance limits are influenced by several factors, most significantly the propagation environment and user mobility pattern. For example, the indoor radio channels typically support higher data rates with better reliability than does the outdoor channel used by persons moving rapidly.

Electromagnetic signals can be characterized by the features of the waveform: amplitude (the power, or magnitude, of the signal); phase (the timing of the peak or trough of the signal variations); and frequency (the number of repetitions of the signal per second).[2] The effects of the wireless channel on the received signal power are typically divided into large-scale and small-scale effects. Large-scale effects involve the variation of the mean received signal power over large distances relative to the signal wavelength, whereas small-scale effects involve the fluctuations of the received signal power over distances commensurate with the wavelength. Path loss effects are noticeable over large distances (i.e., distances on the order of 100 m or more). Signal power variations due to obstacles such as building or terrain features are observable over distances that are proportional to the length of the obstructing object. Very rapid variations result from multipath reflections, which are copies of the transmitted signal that reflect or diffract off surrounding objects before arriving by different paths at the receiver. These reflections arrive at a receiver later than the nonreflected signal path and are often shifted in phase as well. The multipath reflections either reinforce or cancel each other and the nonreflected signal path depending on the exact position of the receiver (if moving) or the transmitter (if moving). The overall effects of multipath propagation involving a moving terminal are rapid variation in the received signal power and nonuniform distortion of the frequency components of the signal.

The first four subsections below discuss path loss, fading, and various sources of interference as they apply to the path between two terrestrial RF devices. The fifth subsection details the characteristics of satellite RF links.

2.1.1.1 Path Loss

Path loss is equal to the received power divided by the transmitted power, and this loss is a function of the transmitter-receiver separation. For a given transmit power, a path loss model[3] predicts the received power level at some distance from the transmitter. The simplest model for path loss, which captures the key characteristics for most channels, is an exponential relationship: The received signal power is proportional to the transmit power and inversely proportional to the square of the transmission frequency and the transmitter-receiver distance raised to the power of a "path loss exponent."[4] In free space the path loss exponent is 2, whereas for typical outdoor environments it ranges from 3 to 5. In environments with dense buildings or trees, path loss exponents can exceed 8. Thus, systems designed for typical suburban or low-density urban outdoor environments require much higher transmit power to achieve the same desired performance in a dense jungle or downtown area packed with tall buildings.

The BER of a wireless link is determined by the received signal power, noise introduced by the receiver hardware, interference, and channel characteristics. The noise is typically proportional to the RF bandwidth. For the exponential path loss model just described, the received signal-to-noise ratio (SNR) is the product of the transmit power and path loss, divided by the noise power. The SNR required for faithful reception depends on the communications technique used, the channel characteristics, and the required BER. Because path loss affects the received SNR, path loss imposes limits on the data rate and signal range for a given BER. In general, for a given BER, high-data-rate applications typically require more transmit power or have a smaller coverage range (sometimes both) than do low-data-rate applications. For example, given a transmit power of 1 W, a transmit frequency of 1 GHz, and an omnidirectional antenna, the transfer of data through free space (for which the path loss exponent is 2) at 1 Mbps and 10^{-7} BER can be accomplished between radios that are 728 m apart, whereas in a jungle (for which the path loss exponent is 10) the range can be as low as 4 m.

2.1.1.2 *Shadow Fading*

A received signal is often blocked by hills or buildings outdoors and furniture or walls indoors. The received signal power is in fact a random variable that depends on the number and dielectric properties of the obstructing objects. Signal variation due to these obstructions is called shadow fading. Measurements have shown that the power, measured in decibels (dB), of signals subject to shadow fading exhibits a Gaussian (i.e., normal) distribution, a pattern referred to as log-normal shadowing. The random attenuation of shadow fading changes as the mobile unit moves past or around the obstructing object. Because the signal coverage is not uniform even at equal distances from the transmitter, the transmit power needs to be increased to ensure that the received-SNR requirements are met uniformly throughout the coverage region. The power increase imposes additional burdens on the transmitter battery and can cause interference for other users of the same frequency band.

2.1.1.3 *Small-Scale (Multipath) Fading*

Small-scale fading is caused by interference between multiple versions of the signal that arrive at the receiver at different times. Multipath can be helpful if the signals add constructively to produce a higher power (a random event), but more often it results in harmful interference. The overall effect is a standing wave pattern of the received signal power. Harmful interference can cause the received signal power to drop by a factor of 1,000 below its average value at nulls in the standing wave pattern. Moreover, for practical speeds of wireless terminals, the changes in signal power are extremely rapid: At a frequency of 900 MHz the signal power changes every 30 centimeters, or every 23 milliseconds if the terminal is moving at 50 km per hour. In many practical environments, these changes are referred to as "Rayleigh fading" because the received signal amplitude conforms to a Rayleigh probability density function.

Signal fading can be characterized by determining the delay spread of the fading relative to the signal bandwidth. The delay spread is defined as the time delay between the direct-path signal component and the component that takes the longest path from the transmitter to the receiver. Because the delay spread is a random variable, it is often characterized by its standard deviation, called the root mean square (RMS) delay spread of the channel. If the product of the RMS delay spread and the signal bandwidth is much less than 1, then the fading is called flat fading. In this case the received signal envelope has a random amplitude and phase (commonly described by a Rayleigh distribution), but there is no additional signal distortion.

When the product of the RMS delay spread and signal bandwidth is greater than 1, the fading becomes frequency selective. Frequency-selective fading introduces self-interference because the delay spread is so large that multipath reflections corresponding to a given bit transmission arrive at the receiver simultaneously with subsequent data bits. This intersymbol interference (ISI) establishes an "error floor" in the received bits that cannot be reduced by an increase in signal power because doing so also increases the self-interference. Without compensation, the ISI forces a reduction in the data rate such that the product of the RMS delay spread and signal bandwidth is less than 0.1. For a 10^{-3} BER and a rural environment, the delay spread is approximately 25 microseconds and the corresponding maximum data rate is only 8 kbps; the data rates for lower BERs are even more limited. Some form of compensation, either signal processing or sophisticated antenna design, clearly is needed to achieve high-rate data transmission in the presence of ISI. These techniques impose additional complexity and power requirements on the receiver.

Movement of a receiver relative to the transmitter (or vice versa) causes the received signal to be frequency shifted relative to the transmitted signal. The frequency shift, or Doppler frequency, is proportional to the mobile velocity and the frequency of the transmitted signal. For a transmitted signal frequency of 900 MHz and a receiver or transmitter speed of 96 km per hour, the Doppler frequency is roughly 80 Hz. This Doppler shift creates an irreducible error floor for noncoherent detection techniques (which use the previous bit to obtain a phase reference for the current bit). In general the irreducible BER is not a problem when data are transmitted at high speed (faster than 1 Mbps), but it is an issue for moderate-rate (slower than 100 kbps) data applications.

In general the signal changes slowly with time because of path loss, more quickly because of shadow fading, and very quickly because of multipath flat fading; all of these effects are simultaneously superimposed on the transmitted signal. As noted above, the shadow fading needs to be addressed by an increase in transmit power. The deep fades in signal power caused by flat fading also need to be counterbalanced by an increase in transmit power or some other approach (see Section 2.1.5.1). Otherwise the transmitted signal typically exhibits bursts of errors that are difficult to correct.

2.1.1.4 Interference

Users of wireless communications systems can experience interference from various sources. One source is frequency reuse, a popular technique for increasing the number of users in a given region who can be supported by a particular set of frequencies. Cellular systems reuse frequencies at spatially separated locations, taking advantage of the falloff in received signal power with distance (which is indicated by the path loss model). The downside of frequency reuse is the introduction of co-channel interference (see Section 2.2.1.1), which increases the noise floor and degrades performance.

Other sources of interference include adjacent channels and narrow bands of problem frequencies. Adjacent-channel interference can be mitigated by the introduction of guard channels between users, although this technique consumes bandwidth. Narrowband interference can be removed by notch filters or spread-spectrum techniques. Notch filters are simple devices that block the band of frequencies containing the interference; these devices are effective only if the specific frequencies of concern are known. Spread-spectrum techniques (see Section 2.1.5.2), which spread a signal across a larger band of frequencies than is required for normal transmission, can reduce the effect of interference and hostile jamming signals.

2.1.1.5 *Satellite Channels*

Satellite channels (the link between a receiver or transmitter on Earth and an orbiting receiver or transmitter) have inherent advantages over terrestrial radio channels. Multipath fading is rare because a signal propagating skyward does not experience much reflection from surrounding objects (except in downtown areas with densely packed buildings). Moreover, most satellite systems operate in the gigahertz frequency range, allowing for large-bandwidth communication links that support very high bit rates.

The primary limitation of satellite channels is very high path loss, which generally follows the formula described earlier in this chapter. For satellites the path loss exponent is 2. Because satellites operate at high frequencies and the path distance is long (500 to 2,000 km for a LEO satellite), much higher transmit power is needed than is the case for terrestrial systems operating at the same data rate. Satellite signals are also subject to attenuation by Earth's atmosphere. The effects are especially adverse at frequencies above 10 GHz, where oxygen and water vapor, rain, clouds, fog, and scintillation cause random variations in signal amplitude, phase, polarization, and angle of arrival (similar to the adverse effects of multipath fading in terrestrial propagation). Satellite systems compensate somewhat for the large path loss and adverse atmospheric effects by using high-gain directional antennas to boost the received power.

2.1.2 Capacity Limits of Wireless Channels

The pioneering work of Claude Shannon determined the total capacity limits for simple wired and wireless channel models: These limits established an upper bound on the maximum spectral link efficiency, measured as the data rate per unit of bandwidth as a function of the received SNR. For a channel without fading, ISI, or Doppler shift, this maximum bandwidth efficiency was identified by Shannon to be the logarithm of the term [SNR + 1] (Shannon, 1949).

Determining the capacity limits of wireless channels with all the impairments outlined in the previous section is quite challenging. A relatively simple lower bound for a channel capacity that varies over time is the Shannon capacity under the worst-case propagation conditions. This is often a good bound to apply in practice because many communication links are designed to have acceptable performance even under the worst conditions. However, this design wastes resources because typical operating conditions are generally much better than the worst-case scenario. For channels that exhibit shadow fading or multipath fading, the channel capacity under worst-case fading conditions is close to zero. The capacity of these fading channels increases greatly when the data rate, power, and transmission are adapted using sophisticated modulation techniques, which are discussed in the next section. As measured by spectral link efficiency, these adaptive techniques in both Rayleigh fading and log-normal-shadowed channels can support much higher data rates than are typical in today's wireless systems. For example, typical digital voice systems deliver 8 kbps in a 30-kHz channel, which corresponds to a spectral link efficiency of 8/30, far less than 1. If this channel experiences Rayleigh fading, then an SNR of approximately 30 dB is required. At this SNR, a spectral link efficiency of approximately 8 can be achieved in Rayleigh fading by using adaptive techniques—a 30-fold improvement over the typical voice system of today (see Figure 2-1).[5]

2.1.3 Modulation

Modulation is the process of encoding information into the amplitude, phase, and/or frequency of a transmitted signal (Ziemer and Tranter, 1995). This encoding process affects the bandwidth of the transmitted signal and its robustness under impaired channel conditions. In the case of bandwidth-limited channels, digital modulation techniques encode several bits into one symbol. The rate of symbol transmission determines the bandwidth of the transmitted signal: the larger the number of bits encoded per symbol, the more efficient the use of bandwidth but the greater the power requirement for a given BER in the presence of noise.

Modulation techniques fall into two categories: linear and nonlinear. In general, linear modulation techniques use less bandwidth than do nonlinear techniques. However, linear modulation techniques also tend to produce large fluctuations in signal amplitude. This is a disadvantage when using nonlinear amplifiers such as class C amplifiers (the least expensive, most readily available, and most power-efficient amplifiers), because they distort linear modulation signals. Thus, the bandwidth efficiency of linear modulation is generally obtained at the expense of the additional power needed for very linear amplifiers (and reduced battery life).

2.1.4 Channel Coding and Link-Layer Retransmission

Channel coding improves performance by adding redundant bits in the transmitted bit stream that are used by the receiver to correct errors introduced by the channel, thus reducing the average BER. This approach enables a reduction in the transmit power required to achieve a target BER. Conventional forward-error-correction (FEC) codes, which re-

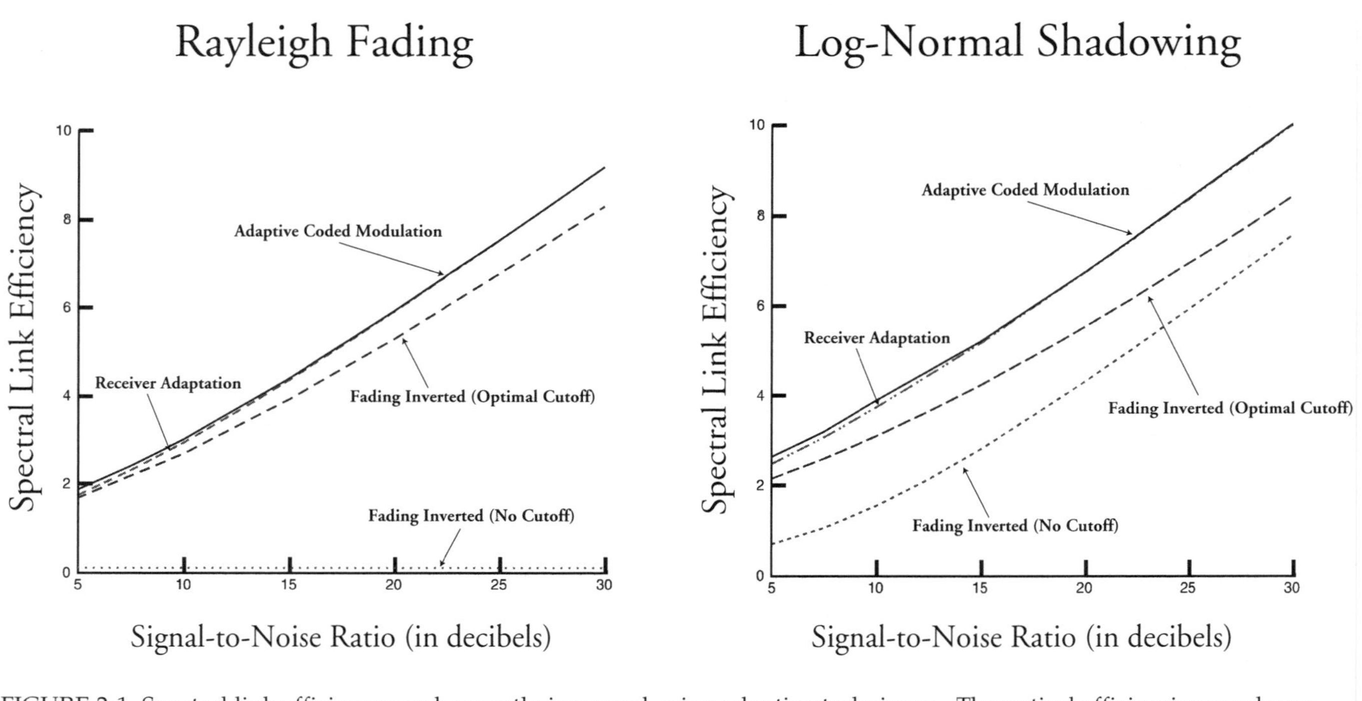

FIGURE 2-1 Spectral link efficiency can be greatly increased using adaptive techniques. Theoretical efficiencies are shown.

duce the required transmit power for a given BER at the expense of increased signal bandwidth or a reduced data rate (Lin and Costello, 1983), use block or convolutional code designs. Block codes add parity bits to blocks of messages. Convolutional codes map a continuous sequence of information bits onto a continuous sequence of encoded bits. Trellis codes combine channel code design and modulation to reduce the BER without bandwidth expansion or rate reduction (Ungerboeck, 1982). More recent advances in coding technology, such as Turbo codes (Berrou et al., 1993), exhibit superior error-correction performance, although they are generally very complex and impose large delays on end-to-end transmission, drawbacks that make them unsuitable for many wireless applications.

Another way to reduce the link errors prevalent in wireless systems is to implement link retransmission (part of the protocol known as automatic repeat request, or ARQ). The data are encoded with a checksum (the sum of the 1s and 0s in the transmitted digitized data), which the receiver compares to the data received; if the data are corrupted, then the receiver requests a retransmission. Link-layer retransmission wastes system resources because of the added power requirements and interference with other users. In addition, retransmission schemes can result in the delivery of data in the wrong order: When a block is lost on the link because of an error burst, a subsequent block is likely to be sent and received before the lost block is sent again. This phenomenon triggers duplicate acknowledgments and end-to-end retransmissions at the transport layer, further burdening the network. (When countermeasures for fading are used to reduce link errors, as discussed in the next section, the problems introduced by retransmission are similarly reduced.) Even so, ARQ is the only alternative in many cases because FEC is not sufficient in applications with stringent BER requirements. Some link-layer schemes, such as asymmetric reliable mobile access in link layer (AIRMAIL; Ayonoglu et al., 1995), can avert out-of-order delivery to higher-protocol layers, but this approach increases delays and variability in the interpacket delivery times.

2.1.5 Countermeasures for Fading

Numerous signal processing and design techniques have been developed to counter the effects of fading on the wireless channel. Of particular interest are countermeasures for the two types of small-scale fading described above: flat fading and frequency-selective fading.

2.1.5.1 Flat-Fading Countermeasures

The random variation in received signal power caused by multipath flat fading results in a very large increase in BER. For example, to maintain an average BER of 10^{-3} (a typical requirement for point-to-point voice systems at the link level) using binary phase-shift key modulation, 60 times more power is required than would be in the absence of flat fading. The difference in required power is even larger at the much lower BER required for data transmission. It follows that the required transmit power can be significantly reduced by combating the effects of flat fading. The most common flat-fading countermeasures are diversity, coding and interleaving, and adaptive modulation. Spread-spectrum techniques also mitigate fading effects (see Section 2.1.5.2).

In diversity, several separate, independently fading signal paths are established between the transmitter and receiver and the resulting received signals are combined. Because there is a low probability of separate fading paths experiencing deep fades simultaneously, the signal obtained by combining several such paths is unlikely to experience large power variations. Independent fading paths can be achieved by separating the signal in time, frequency, space, or polarization. Time and frequency diversity are spectrally inefficient because information is duplicated; polarization diversity is of limited effectiveness because only two independent fading paths (corresponding to horizontal and vertical polarization) can be created. That leaves space diversity as the most efficient of these techniques. Independent fading paths in space are obtained using an antenna array, in which each element receives a separate path. Multiple antenna elements are mounted at the receiver with a separation greater than or equal to half the signal wavelength (Yacoub, 1993). Almost all of the multipath variation is removed by first creating and then later combining four independent paths, with each path weighted by its received signal power. Because the wavelength is inversely proportional to frequency, antenna arrays can be mounted on handheld units when using superhigh frequencies (above 10 GHz) but not when using frequencies below the 1-GHz range.

Coding and interleaving can also be used to combat flat fading. Coding and interleaving involves the spreading of a burst error over many "code words." If the errors are sufficiently spread out that each code word has at most one error, then these errors can be corrected easily. This technique results in time diversity without the need for repeat transmissions. However, error-correcting codes typically result in the loss of spectral link efficiency. The cost of coding and interleaving—increased delay and complexity of the interleaver—can be large if the fading rate is slow relative to the data rate, as is typically the case for high-speed data. For example, at a Doppler frequency of 10 Hz and a bit rate of 10 Mbps, an error burst will last for approximately 300,000 bits, and so the interleaver needs to be large enough to handle at least that much data.

In general, flat fading causes bit errors to occur in bursts corresponding to the times when the channel is in a deep fade. Channel codes (discussed in Section 2.1.4) are best suited for correcting one or two simultaneous errors; code performance deteriorates rapidly when errors occur in large bursts.

The third type of countermeasure is adaptive modulation. In theory, the receiver can make an estimate of the channel conditions and send it back to the transmitter, which can then adapt its transmission scheme as appropriate. Adaptation to signal fading enables adjustments in the power level and data rate to take advantage of the favorable conditions, saving more than 20 dB of power. But most modulation and coding techniques do not enable sufficiently rapid adaptation to typical fading conditions. If the channel is changing more rapidly than the rate at which condition estimates are fed back to the transmitter, then adaptive techniques perform poorly. Another drawback is the additional complexity required in the transmitter and receiver to carry out all the requisite steps. Finally, the channel estimate needs to be relayed to the transmitter on a feedback path, which occupies a small amount of bandwidth on the return channel.

2.1.5.2 Countermeasures for Frequency-Selective Fading

Techniques for combating the ISI delay spread of a frequency-selective fading channel fall into two categories: signal processing (at the transmitter or receiver) and antenna solutions. Transmitter-based signal processing techniques, including equalization, multicarrier modulation, and spread spectrum, can make the signal less sensitive to delay spread. Antenna solutions, including distributed antenna systems, small cells, directive beams, and "smart" antennas, change the propagation environment to reduce or eliminate delay spread.

The goal of equalization is to invert the effects of the channel or cancel the ISI. Channel inversion, or linear equalization, can be achieved by passing the received signal through a filter with a frequency response that is the inverse of the channel frequency response (the channel being the original "filter" for the transmitted signals). This process neutralizes the effects of the channel. Although linear equalization can be implemented using relatively simple hardware, the technique has drawbacks in that noise and interference are also passed through the inverse filter. If the channel frequency response is small anywhere within the signal bandwidth, then the noise and interference components at those frequencies are amplified. Thus, on channels with deep spectral nulls, a linear equalizer enhances the noise, resulting in poor performance on frequency-selective fading channels.

A more effective technique is the nonlinear decision-feedback equalizer (DFE), which determines the ISI from previously detected symbols and subtracts it from the incoming symbols. The DFE does not enhance noise because it estimates the channel frequency response rather than inverting it. On frequency-selective fading channels the DFE has a much lower probability of error than does a linear equalizer but also slightly higher complexity. The main drawback of the DFE is the chance of error propagation: If a symbol is detected incorrectly then the associated ISI is still subtracted from subsequent symbols, possibly causing errors in these symbols as well. Moreover, because of decoding delays, the ISI estimates cannot benefit from error-correction coding. Therefore, a DFE can be used only on channels where the probability of error without coding is reasonably low. In addition, because the ISI at low data rates is small, the DFE does not yield substantial BER improvement for data rates much less than 100 kbps. In general, equalizers (especially the DFE) are most beneficial at high data rates, when the product of RMS delay spread and the data rate is much greater than 1. The BER can be improved by as much as 2 to 3 orders of magnitude depending on the data rate and SNR. In indoor environments, 20 Mbps can be achieved at a BER of 10^{-3} using a DFE (Pahlavan et al., 1993). The achievable rates on outdoor channels are generally much less because delay spreads are much greater and there are other channel impairments. Commercial outdoor wireless systems using delay-spread countermeasures currently achieve on the order of tens to hundreds of kilobits per second, depending on the available bandwidth.

The use of equalizer techniques requires continuous, accurate estimates of the channel frequency response, usually obtained with finite-impulse-response (FIR) filters in the receiver. The number of filter delay elements (or equalizer taps used to track variations in the channel) is proportional to the delay spread. Updates are needed at the Doppler rate. To assist in the estimation process, the transmitter sends training sequences that have known characteristics. Because they consume bandwidth, training sequences need to be as short as possible to maximize spectral link efficiency. The trade-off is that short training sequences require rapid estimation of the channel and, typically, increased signal-processing complexity.

Multicarrier modulation is another technique that compensates for delay spread. The transmission bandwidth is divided into subchannels and the information bits are divided into an equal number of streams, which are transmitted in parallel. Each stream is used to modulate one of the subchannels. Ideally, the subchannel bandwidths are narrow enough that the fading on each subchannel is flat as opposed to frequency selective, thereby eliminating ISI. The simplest approach is to implement nonoverlapping subchannels, but spectral link efficiency can be increased by overlapping the subchannels in such a way that they can be separated at the receiver. This is called orthogonal frequency division multiplexing, which can be implemented efficiently using the fast Fourier transform (invertible mapping from the time domain to the frequency domain) to separate the subchannels in the receiver. In this case the entire signal bandwidth experiences frequency-selective fading because wideband channels tend to have different fading characteristics at different frequencies, and so some of the subchannels will have weak SNRs. Their performance can be improved by coding across subchannels, frequency equalization, or adding more bits in subchannels with high SNRs. Multicarrier modulation offers an advantage in that less training is required for frequency equalization than for time equalization. However, time-varying fading, frequency offset, and timing mismatch impair the separation of the subchannels, resulting in self-interference. Moreover, multicarrier signals tend to have a large peak-to-average signal-power ratio, which severely degrades the power efficiency when nonlinear amplifiers are used.

Spread-spectrum techniques increase the signal bandwidth—beyond what is needed to transmit the information—to reduce the effects of flat fading, ISI, and narrowband interference. Each channel is spread over the larger bandwidth by a pseudo-noise sequence, which is used by receivers to invert the spreading operation and recover the original data. Spread-spectrum techniques first achieved widespread use in military applications because they "hide" the signal below the noise floor during transmission, reduce the effects of narrowband jamming, and reduce multipath fading. There are two common forms of this technique: direct sequence, in which the data sequence is multiplied by the pseudo-noise sequence, and frequency hopping, in which the carrier frequency is varied by the sequence.

During the demodulation process, multipath signal components and interference are reduced in two stages: First the spectrum-spreading modulation is removed, and then the remaining signal is demodulated using conventional frequency- or phase-shift techniques to obtain the original data signal. In direct-sequence systems, the received signal is multiplied with an exact copy of the code sequence, perfectly synchronized in time. When narrowband interference and delayed multipath signal components are multiplied by the spreading sequence, their signal power is spread over the bandwidth of the spread-spectrum code. A narrowband filter can be used in the demodulator to remove most of their power. Alternatively, a RAKE receiver can be used to combine all multipath components coherently.[6] A RAKE is a means of implementing diversity (see Section 2.1.5.1).

2.1.6 Channel Access

In many modern wireless systems, multiple users share the same bandwidth, creating a need for a protocol that ensures efficient, equitable channel access. Wireless-channel access issues are complicated by the variability and statistical nature of user traffic: Voice traffic typically requires a 40 percent duty cycle (i.e., the channel is used 40 percent of the time), whereas data traffic tends to come in bursts with a much lower duty cycle. All traffic generally varies depending on how many transmitters are active. In addition, many new applications do not exhibit the symmetric two-way flow of voice data that is characteristic of standard telephone service. In typical surfing of the World Wide Web, for example, 100 to 1,000 times more data flows to the user than from the user. This variability and asymmetry are creating a need for new access strategies for digital integrated networks.

Channel sharing through fixed-allocation, demand-assigned, or random-allocation modes is called multiple access.[7] The three basic multiple-access techniques—FDMA, TDMA, and CDMA (all introduced in Chapter 1)—can be implemented in any of the three modes.

2.1.6.1 Fixed-Allocation Multiple Access

Fixed-allocation multiple-access techniques assign dedicated channels to multiple users through some type of channel resource division. The assignments are made by a protocol for the duration of a single transmission.[8] In FDMA the total system bandwidth is divided into channels that are allocated to the different users. In TDMA time is divided into orthogonal slots that are allocated to different users. In CDMA (which is the same as direct sequence spread spectrum) time and bandwidth are used simultaneously by different users, modulated by different spreading signals, or codes. The spreading codes allow receivers to separate the signal of interest from the CDMA channel. The three primary competing U.S. standards for cellular and personal-communications multiple access are mixed FDMA/TDMA with three time slots per frequency channel (IS-54), mixed FDMA/TDMA with eight time slots per frequency channel (GSM), and CDMA (IS-95).

The debates over multiple access among standards committees and equipment providers have led to numerous analytical studies claiming the superiority of one technique or another (e.g., Gilhousen et al., 1991; Gundmundson et al., 1992; Baier et al., 1996). However, there is no widespread agreement as to which access technique is the best. Theoretical analysis indicates that under heavy traffic conditions CDMA combined with some form of detecting all users simultaneously[9] (using knowledge of all spreading codes to eliminate interference) results in higher spectral efficiency than does TDMA or FDMA (Gallager, 1985; Goldsmith, 1997). Without simultaneous detection and in the absence of fading, TDMA and FDMA are more spectrally efficient than is CDMA.[10] The spread spectrum gives CDMA the advantage of soft capacity (there is no absolute limit on the number of users), but performance is degraded in proportion to any increase in users on the system. The TDMA and FDMA techniques place hard limits on the number of users sharing a given bandwidth because each time or frequency slot can support a maximum of one user (less than one if multiple slots are assigned to the same user). In general FDMA is the simplest technique to implement, TDMA is slightly more complex because of the requirement for time synchronization among all the users, and CDMA is the most complex because of the need for code synchronization. Another consideration with respect to CDMA is the need for stringent power control to prevent the "near-far problem," which arises when signals from mobile units close to the base station overwhelm those of units farther away. Such control is difficult to maintain in a fading environment and is one of the major challenges of spread-spectrum multiple access.

Fixed-allocation multiple-access techniques are inefficient for many voice and data applications because the variability in traffic from a single transmitter limits throughput on dedicated channels. For example, a single channel in a two-way voice conversation usually occupies less than half of the available bandwidth; for many data applications the traffic is even more intermittent. Cellular and satellite systems generally serve a slowly changing set of active terminals with a relatively fixed traffic pattern. The inability to predict terminal traffic requirements accurately and the need to handle a dynamic set of active terminals create a need for more flexible forms of multiple access.

2.1.6.2 *Demand-Assigned Multiple Access*

One method of providing flexibility is the assignment of network channels to remote terminals on demand. In these systems a common signaling channel is assigned to handle requests from transmitters for network capacity. Demand-assigned multiple access (DAMA) is very efficient as long as the "overhead" traffic required to assign channels is a small percentage of the message traffic and as long as the message traffic is fairly steady. Otherwise two types of problems can arise. First there is a set-up delay, or latency period. For transmissions of sufficient length this is not a serious limitation, but for networks with a considerable amount of short, interactive messages the delay and overhead of each message make demand-based assignment impractical.[11] Second, the transmission of requests on the signaling channel is not possible when the network is overloaded and the transmitter cannot communicate with the hub station, effectively shifting the multiple-access problem from the data channel to the signaling channel.

2.1.6.3 *Random Access*

When networks serve a wide variety of data rates and the traffic consists of small messages that are roughly the same size as the overhead messages of the access protocol, DAMA is not an efficient use of channel resources. In these cases a connection-free protocol, such as random-access CDMA or ALOHA, is preferable. The random-access CDMA approach requires complex receivers that can demodulate all possible spreading codes. In ALOHA random access, channel packets are stored at each terminal and transmitted over a common channel to the hub station; no attempt is made to synchronize transmissions from the various users. This technique has high reliability in moderate network traffic, but the probability of a collision between packets from different users increases with the traffic. Therefore, such channels are usually sized to operate at about 10 percent of the peak data rate.[12]

Conventional ALOHA is a narrowband technology. In wireless networks, ALOHA channels rarely operate at more than 10 kbps or 20 kbps in terrestrial systems and 56 kbps in satellite systems. Because ALOHA can be viewed as a random-access version of TDMA, the same peak-power constraints that limit a TDMA channel also apply to conventional ALOHA channels. Recent work on spread ALOHA, a combination of ALOHA and spread-spectrum transmission, shows how these limits can be overcome for high-data-rate applications.

Throughput is not necessarily the most appropriate performance measure for a multiple-access channel. In the case of power-limited satellite channels or battery-operated transmitters, access efficiency is a more appropriate measure. The access efficiency of an ALOHA random-access channel is the ratio of spectral link efficiency using the ALOHA protocol to the spectral link efficiency of a continuously transmitting channel with the same average power and total bandwidth. The access efficiency of an ALOHA channel approaches 1 (meaning no restrictions are needed on throughput) when most users are idle and transactions are brief, as can be the case for some data communications systems. In other words, this technique offers the highest throughput of any random-access protocol under these conditions.

It is much easier to design an access protocol for a single type of network traffic rather than for a range of traffic types. All the major digital cellular standards designed for voice applications, use a DAMA architecture with some form of ALOHA request channel superimposed over a TDMA or CDMA channel structure. The resulting throughput is adequate for voice applications, but when a network handles data as well as voice the connection-oriented architecture limits the channel throughput. It is difficult to size channels that are assigned on demand for a wide and unpredictable range of user data rates. New, highly flexible random-access structures will probably be needed to enable the seamless integration of data services within a voice network as promised in some new personal communications networks.

2.2 NETWORK ISSUES

2.2.1 Architecture

The choice of an architecture for a two-way wireless network involves numerous issues dealing with the most fundamental aspects of network design. The primary issue is whether to use a peer-to-peer or a base-station-oriented network configuration. In a peer-to-peer architecture, communication flows directly among the nodes in the network and the end-to-end process consists of one or more individual communication links. In a base-station-oriented architecture, communication flows from network nodes to a single central hub.

The choice of a peer-to-peer or base-station-oriented architecture depends on many factors. Peer-to-peer architectures are more reconfigurable and do not necessarily have a single point of failure, enabling a more dynamic topology. The multiple hops in the typical end-to-end link offer the advantage of extended communication range, but if one of the nodes fails then the localized link path needs to be reestablished. Base-station-oriented architectures tend to be more reliable because there is only one hop between the network node and central hub. In addition, this design tends to be more cost-efficient because centralized functions at the hub station can control access, routing, and resource allocation. Another problem with peer-to-peer architecture is the significant co-site interference that arises for multiple users in close proximity to each other—a problem that can be averted in a base-station-oriented architecture by the coordinated use of transmission frequencies or time slots.

The wireless base-station-oriented architecture is exemplified by cellular telephone systems, whereas the most common peer-to-peer architecture for wireless systems is a multihop packet radio. Fundamental differences between the two types of systems are indicated in Table 2-1.

TABLE 2-1 Comparison of Cellular and Multihop Packet Radio Architectures

Feature	Cellular System	Multihop Packet Radio
Topology	Static	Dynamic
Number of hops	One	Multiple
Network control	Centralized	Distributed
Link distance	Fixed (by cell size)	Variable ("over the horizon")

2.2.1.1 Cellular System Design

One of the biggest challenges in providing multimedia wireless services is to maximize efficient use of the limited available bandwidth. Cellular systems, which exploit the falloff in power at increased distances, reuse the same frequency channel at spatially separated locations. Frequency reuse increases spectral efficiency but introduces co-channel interference, which affects the achievable BER and data rate of each user. The interference is small if the users operating at the same frequency are far enough apart; however, area spectral efficiency (i.e., the data rate per unit bandwidth per unit area) is maximized by packing the users as close together as possible. Thus, good cellular system design places users that share the same channel at a separation distance such that the co-channel interference is just below the maximum tolerable level for the required BER and data rate. Because co-channel interference is subject to shadowing and multipath fading, the design of a static cellular system needs to assume worst-case propagation conditions in determining this separation distance. System performance can be improved through dynamic resource allocation, which involves allocating power and bandwidth based on propagation conditions, user demands, and system traffic; however, the increases in spectral and power efficiency are achieved at the price of increased system complexity.

In cellular systems, a given geographical area such as a city is divided into nonoverlapping cells (see Figure 2-2) and different frequencies are assigned to the cells. For FDMA and TDMA, cells using the same frequency are spatially separated such that their mutual interference is tolerable. Frequency reuse in FDMA and TDMA systems introduces co-channel interference in all cells using the same channel. In CDMA, interference can be introduced from all users within the same cell as well as from users in other cells.

Regardless of the multiple-access technique used, area spectral efficiency is maximized when the system is interference limited—that is,

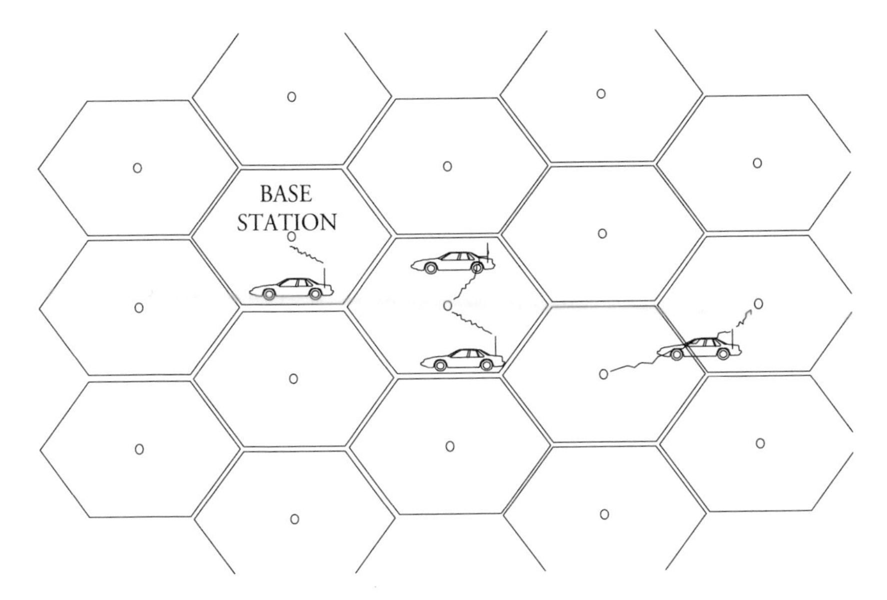

FIGURE 2-2 In cellular systems each cell has a central hub (the base-station-oriented design) and is assigned different frequencies, time slots, or codes, depending on how users access the system.

when the receiver noise power is much less than the interference power and can be ignored. The received SNR for each user is determined by the amount of interference at the receiver; if the system is not interference limited, then spectral efficiency could be further increased by allowing additional users on the system or reusing the frequencies at reduced distances.

The transmitter in each cell is connected to a base station and switching office, which allocates channels and controls power. In analog cellular systems, the switching office also coordinates handoffs to neighboring cells when a mobile terminal traverses a cell boundary. In digital cellular systems and low-tier systems, base stations and terminals play a more active role in coordinating handoffs. The spectral efficiency can be increased by dividing each existing cell into several smaller cells because more users can then be accommodated in the system. However, reducing the cell size increases the rate at which handoffs occur, sometimes affecting higher-level protocols. In general, if the rate of handoff increases, then the rate of call dropping will increase proportionally. Routing is also more dynamic with small cells because routes need to be reestablished whenever a handoff takes place.

2.2.1.2 *Packet Radio System Design*

A packet radio system provides communications to fixed and mobile network nodes that use radios to form the physical links (Lauer, 1995; Leiner et al., 1997). The earliest such systems were the ALOHANET, which operated at the University of Hawaii in the early 1970s, and the DARPA PRNet, a multihop peer-to-peer packet network that operated in the late 1970s and early 1980s. Commercial packet radio networks have been built around single-hop base-station-oriented architectures, as in the Ardis or Mobitex systems, and multihop peer-to-peer architectures, as in the Metricom system (see Chapter 1, Section 1.6). These networks can be constructed with fixed-location infrastructure elements (as in Metricom) or can achieve connectivity in a completely ad hoc manner. In general, multihop ad hoc packet-radio networks can be set up, deployed, and redeployed rapidly. These characteristics are important to military operations. However, multihop ad hoc packet radio networks can pose difficulties in defense applications, because a peer-to-peer architecture does not correspond to the military command structure.

Many of the challenges in packet radio system design are the same as those for any wide-area wireless communications system. These issues include how best to deal with the fading characteristics of RF propagation and whether to use a random or reservation access strategy. Packet radio networks also pose special challenges related to the dynamic nature of the network topology. The terrain, distance between nodes, antenna height, and antenna directionality all influence whether network connectivity can be established and maintained. Physical connectivity in ad hoc packet radios is more complex than in cellular systems because cell sites cannot be surveyed in advance and may be situated in locations that are difficult to access. Furthermore, it is not economical (commercially at least) to use large antennas or extensive antenna processing and directionality at each peer-to-peer node; the network nodes are all more or less identical, highly portable, and always moving, although additional repeaters can be added within the system to improve performance. Repeaters demodulate packets, remodulate them, and send them again.

Military packet radio systems typically operate at lower frequencies than do cellular systems so as to cover large areas within the battlefield. Active interference needs to be considered in system design, and transmitter power is chosen not only to ensure successful reception at the receiver but also to hide the network from adversaries. Military packet radio systems make extensive use of spread-spectrum methods for channel access and in general require a higher degree of flexibility in coding schemes than do commercial systems. Preamble spreading codes (simple versions of the data spreading codes used for synchronization or header information) may be different from the codes used during the data portion of the packet, and codes can be changed on a bit-by-bit basis to reduce the probability of interference (a feature of second-generation DARPA packet radios). All transmitters use either a common preamble code or a receiver-directed preamble code that directs the transmission to a single node that is tuned to the specific code. The latter approach makes it possible for multiple packets to be in the air yet have a low probability of interference.

2.2.2 Physical Resource Allocation

Any system using a fixed assignment of network resources needs to be designed based on worst-case signal propagation and interference assumptions. A more efficient strategy is dynamic resource allocation, in which channels, data rates, and power levels are assigned depending on the current interference, propagation, and traffic conditions. For cellular systems, dynamic resource allocation includes assignment of channels to base stations. Dynamic channel allocation in cellular systems improves channel efficiency by a factor of two or more, even when using simple algorithms (Katzela and Naghshineh, 1996). However, analyses of dynamic resource allocation to date have been based on fairly simplistic system assumptions, such as fixed traffic intensity, homogenous user demands, fixed reuse constraints, and static channels and users. Little work has been done on resource allocation strategies that consider simultaneous, random variations in traffic, propagation, and user mobility. The extent to which system performance can be improved under realistic conditions remains an open and challenging research problem with respect to both cellular and packet-radio architectures. Previous research has focused primarily on cellular systems; little attention has been devoted to peer-to-peer networks.

An emerging and important research area focuses on reducing the complexity of dynamic resource allocation, particularly in systems with small cells and rapidly changing propagation conditions and user demands. Even under simplistic assumptions of fixed conditions, optimizing channel allocation is highly complex. Current allocation procedures are not easily generalized to incorporate power control, traffic classes (e.g., multimedia), cell handoff, or user priorities. In addition, the efficiency of dynamic resource allocation is most pronounced under light loading conditions. Thus, the optimal dynamic resource allocation strategy is also dependent on traffic conditions.

2.2.3 Interoperability

For elements of a system to communicate, they must be compatible. One way to achieve compatibility is to mandate a "point design" in which all devices conform to the same standard. As described in Chapter 1, this approach was taken for first-generation cellular systems in the United States and in second-generation systems in Europe. As used in this report, the term "interoperability" refers to the capability of network elements that do not conform to the same standard to communicate. Interoperability can be achieved in two ways using different enabling devices: gateways and adapters. In the compatibility context, a gateway is a device that conforms to more than one standard, whereas an adapter translates information formats between two standards. A cable-ready television set is an example of a gateway, and a set-top box is an example of an adapter.

With respect to military wireless communications systems, there will be no convergence to a single technology in the foreseeable future, for many reasons. The number of incompatible systems will remain high, and yet evolving military missions will require increasing communications between individuals and machines using different systems. As a consequence, interoperability among these systems will be essential.

In sophisticated multimedia networks such as the ones required for military operations in the next century, interoperability is necessary at all layers of a communication protocol. The Internet approach to interoperability is a narrow-waist protocol suite with compatibility at the middle layers (TCP and IP) and diversity at higher (i.e., application) and lower (i.e., physical) layers (Computer Science and Telecommunications Board, 1994). Although this approach can be adopted in all types of communications systems, there are several physical-layer problems unique to wireless communications. For example, wireless systems may operate in different frequency bands and use different modulation and coding techniques. Multimode radios are gateways that address these problems. The commercial dual-mode cellular telephone is an example of this type of radio.

The software radio is a promising means of achieving interoperability at the physical layer. Software radio receivers digitize the RF signal and implement most receiver functions by means of software running on general-purpose hardware (see Section 2.4). Similarly, transmitters synthesize waveforms in digital format and convert them to analog prior to amplification. A software radio can be programmed to be compatible with a number of communications systems and provide interoperability across the required data encoding, transmit waveforms and bandwidths, timing, and clock accuracy of the individual modes. Chapter 3 (Section

3.4.3.1) notes the initial success of experimental SpeakEASY radios in implementing gateways between several military waveforms. However, current-generation software radios are limited in terms of the range of radio waveforms they can handle.

The implementation of gateways and/or adapters raises the issue of where they would best be situated in a communications system. There are various possibilities depending on network architecture. In peer-to-peer networks, the terminals need to be capable of implementing all coexisting technologies. In this case any terminal would be capable of communicating with any other terminal within range; the disadvantage would be the added cost, weight, and power drain relative to single-mode terminals. On the other hand, in base-station-oriented networks, it may be possible to concentrate the tasks of interoperability in base stations. This approach has the disadvantage of disabling communications between terminals when base stations are out of service. This issue could be addressed in research on network architectures (see Chapter 4).

2.2.4 Routing and Mobility Management

2.2.4.1 Multihop Routing

The routing of messages through a multihop packet-radio network requires the identification of existing communication links and an assessment of their relative quality. Routing protocols perform these tasks. The best route is the one with the smallest number of hops providing acceptable connectivity; the link quality can be determined by measuring signal strength, SNR, or BER. Poor link quality can be improved to some extent through the use of higher transmission power, wider spreading codes, aggressive hop-by-hop error correction, or retransmission schemes. However, link capacity is also a function of the traffic on nearby links; it may be necessary to route around nodes experiencing heavy congestion.

In general, network topologies vary rapidly in mobile packet radio networks, with links constantly being lost and new ones established. Therefore, the network management component needs to disseminate connectivity information more rapidly than is necessary in wired networks. The network also needs to be able to handle gracefully any network partitions caused by link outages, which are more likely to occur in mobile packet radio networks than in a conventional wired network.

Routing algorithms choose a hop-by-hop path based on information about link connectivity. The simplest scheme is flooding, in which a packet is transmitted on all links from the source to neighboring nodes, which then repeat the process. Flooding is inefficient but can be the best strategy when a network topology changes rapidly. Another scheme, connection-oriented routing, maintains a sequence of hops for communications between a single source and specific destination in the network. Given rapid topology changes, network partitions, and large numbers of nodes, keeping this information updated and available to all nodes is difficult at best. A third scheme is connectionless routing, which requires no knowledge of end-to-end connections. Packets are forwarded toward their destination, with local nodes adapting to changes in network topology.

Connection-oriented and connectionless approaches require that routing information be distributed throughout the network. In small networks this distribution was originally accomplished by a centralized routing server; by now, distributed algorithms with improved scaling behavior have largely replaced centralized servers, especially in large networks. Each node independently determines the best hop in the direction of the destination, and updated routing tables are periodically exchanged among neighboring nodes.

Routing schemes have also been used that combine elements of the centralized and distributed approaches. For very large multihop packet radio networks, such schemes impose a hierarchy on the network topology, hiding changes in the distant parts of the network from local nodes (the next-hop routes to distant network nodes are not likely to change as rapidly as are routes within each cluster). A combined strategy is to use a centralized route server, known as a cluster head, to maintain routes between clusters in the direction of "border radios."

A final routing issue relates to packet forwarding, which is initiated when several transmission attempts fail to deliver a message to the next node. In these cases a node engages in localized rerouting, broadcasting the message to any node that can complete the route. Packet forwarding can cause flooding if multiple nodes hear the request and choose to forward the packet. The process can be optimized by filtering based on overheard traffic: If a node has a packet in its send queue and "hears" the same packet being forwarded from a second node, then the first node assumes that the packet has been sent and removes it from the queue.

2.2.4.2 *Terminal Mobility*

The mobile internetworking routing protocols (Mobile IP) were designed to accommodate the mobility of Internet users. There is some disagreement concerning whether Mobile IP was originally designed for an individual user moving from one fixed location to another (Myles et al., 1995) or for on-the-move wireless operations. In any case, the Internet Engineering Task Force (IETF) has addressed the issue of full mobility, and Mobile IP is now suited for highly mobile users.[13] On the Internet, every node (fixed or mobile) has a unique identifying address, its IP address. Mobile IP has to circumvent the association of IP addresses with specific networks because mobile nodes can attach to and detach from multiple networks as they roam. Changing an IP address on the fly is not always possible. If a node has an open TCP connection when its IP address changes, then the TCP connection will fail. If the node requires an accurate Domain Name System (DNS) entry, then the entry will need to be updated as the address changes, and in today's implementation of DNS such an update can be very slow.

Communications take place between a sender and receiving mobile host (MH). In the Mobile IP specifications, every MH has a home network and its IP address is called the home address. A router called the home agent, which resides in the MH's home network, is responsible for intercepting each packet destined for the home address of a roaming MH. The packet is placed inside another IP packet through a process called "IP-in-IP encapsulation." The source address in the encapsulating packet is that of the home agent. The packet is usually sent "in care of" another agent, the foreign agent, which resides in the network in which the MH is roaming. The packet is sent by conventional IP routing to the foreign agent, where the contents (i.e., the original packets) are removed and delivered to the MH. The MH can transmit information directly to the sender but the sender always directs its own communications to the home network. The MH can also request a locally assigned care-of IP address in its roaming domain by invoking the dynamic host configuration protocol; this address could be used by the home agent directly, eliminating the foreign agent.

When an MH enters a new mobile subnetwork it needs to obtain a care-of address. It can find a foreign agent using a process built on top of the existing Internet control message protocol capabilities for router discovery. Once accepted by the local network, the MH registers its new care-of address with its home agent. All registration attempts need to be carefully authenticated to prevent a malicious user from hijacking the packets simply by furnishing another care-of address. The Mobile IP specifications use a message authentication code (similar to a digital signature) based on a secret key shared by the MH and home agent, typically using a secure one-way function called MD5. Only the MH that knows the secret key can provide the digital signature expected by the home agent. Replay protection is required to prevent a malicious user from falsely registering an MH with a stale care-of address.

The major performance challenge is to circumvent the indirect routing among the sender, home agent, and foreign agent. This path can be eliminated if the sender caches bindings between the MH's home and care-of addresses. The management of these bindings is called route optimization.[14] For example, when a sender first sends a packet to an MH through its home agent, the home agent could send a binding-update message to the original sender. Until the binding expires because of a time-out, the sender can use the care-of address directly. If the MH moves to a new subnetwork, then it can ask its former foreign agent to forward packets to the new care-of address while also alerting senders of that new address.

2.2.4.3 *Wireless Overlay Networks*

No single network technology can simultaneously offer wide-area coverage, high bandwidth, and low latency. In general, networks that span small geographical areas (e.g., LANs) tend to support high bandwidth and low latency, whereas networks that span large geographical areas (e.g., satellite networks) tend to support low bandwidth and high latency. To yield flexible connectivity over wide areas, a wireless internetwork needs to be formed from multiple wide-, medium-, and local-area networks interconnected by wired or wireless segments (Katz and Brewer, 1996). This internetwork is called a wireless overlay network because the WANs are laid on top of the medium- and local-area networks to form a multilayered network hierarchy. A user operating within the LAN enjoys high bandwidth and low latency, but when communicating outside the local coverage area the user accesses a wider-area network within the hierarchy, typically sacrificing some bandwidth or latency in the process.

Future mobile information systems will be built on heterogeneous wireless overlay networks, extending traditional wired and internetworked processing "islands" to hosts on the move over a wide area. Overlay technologies are used in buildings (wireless LANs), in metropolitan areas (packet radio), and regional areas (satellite). The software radio, with its capability to change frequencies and waveforms as needed, is a critical enabling technology for overlay networks.

Handoffs may take place not only "horizontally" within a single network but also "vertically" between overlays. If each overlay network assigns the MH a different IP address, then Mobile IP needs to be extended to correlate all the addresses for one user. Alternatively, the mobile node can treat each new IP address as a new care-of address. The home agent maintains a table of bindings between the home and locally assigned addresses. The applications running on the MH may participate in the choice of route. For example, an application might specify that high-priority traffic traverse an overlay with low latency. Less-critical traffic might travel over higher-latency connections. Signal quality, BER, and packet loss and retransmission need to be considered. Under certain conditions such as the transmission of urgent data, a slow-speed overlay with a strong signal strength might be preferred to one with a higher bandwidth but a weaker signal.

2.2.5 Resource Discovery

New protocols are being developed to support convenient operations by mobile users. One example is the service location protocol, which allows user agents to determine access information for generic network services such as printing, faxing, schedule management, file system access, and backup. A directory agent delivers universal resource locators (URLs) to user agents, which use the URLs to access service agents. New service agents can register or withdraw their URLs as needed. Much of the protocol research is geared toward enabling the identification of directory agents in unfamiliar environments. Other strategies based on modifications to directory services have been proposed as well.

2.2.6 Network Simulation and Modeling Tools

Network performance analysis can take three forms: mathematical analysis, experimental trials, or system simulation. Mathematical analyses can incorporate only a limited range of realistic phenomena, and field trials are expensive as well as difficult to set up under repeatable conditions. Consequently simulation is often the best tool for optimizing system design and predicting performance.

Network-level simulation tools are used to simulate the dynamic behavior of routing, flow, and congestion-mitigation schemes in packet-switched data networks. These tools can model arbitrary network topologies, link-error models, router scheduling algorithms, and traffic.[15] Network simulators have been used to investigate new link-layer algorithms for packet scheduling and retransmission, new mechanisms within routers for determining local congestion conditions and sending this information to transmitters and receivers, and transport-layer algorithms for retransmission and rate control. Performance tools can also help troubleshoot problems in real networks by collecting statistics about the throughput of the various nodes and links. This information can be used to identify bottlenecks and develop remedial strategies such as changing the topology of the network. Debugging tools enable the protocol designer to capture detailed traces of network activity (McCanne and Jacobson, 1993); these tools are invaluable for tracking down errors in protocol implementation.

To model mobile networks accurately, simulators require special features, some of which have yet to be developed. They need to model the nature of errors on the wireless link precisely because errors are not uniformly distributed but rather tend to cluster (Nguyen et al., 1996). They also need to model node mobility, especially in the case of packet radio networks. Existing simulation technology consists of good models of radio propagation at microwave frequencies but only standard teletraffic models and few abstract mobility models. Some proprietary tools integrate geographical modeling, propagation, and cellular networking behavior, but no integrated tools are available commercially to predict the performance of the next generation of wireless technologies, such as smart radios.

Similarly, existing tools can simulate the creation of relatively narrowband waveforms at the transmitter and analyze the effects of radio propagation on the received signal, but they cannot model the antenna radiation or reception properties of a signal that spans more than 1 GHz of spectrum. No existing tool can model the propagation performance of urban, suburban, rural, or free-space radiation of wideband signal-containing components with diverse propagation characteristics. No tool can analyze the effects of the motion of network elements on the received signal's multipath characteristics, such as spectral nulls and Doppler shift over wide bandwidths. No widely available tools allow for the geographical or topological analysis of specific network deployments. Finally, no analysis tool is sophisticated enough to examine the performance of software radios or radio networks in the presence of interference sources common to wideband mobile communications. The evaluation and optimization of mobile wireless networks would be enhanced by the development of sophisticated, flexible models of communications traffic and node mobility.

2.3 END-TO-END SYSTEM DESIGN ISSUES

Most end-to-end system design issues, such as security, design tools, and interoperability with other systems, are relevant to any wireless application (Katz, 1994). However, some end-to-end design issues depend on the application(s) to be supported by the network. For example, videoconferencing is an extremely challenging application for a wireless system because of its high bandwidth requirement and strict constraints on delayed end-to-end transmissions. To support this application, the capability to adapt to channel conditions, perhaps through a slight degradation in image quality, might be built into the end-to-end system protocols. Similarly, many military command-and-control operations require the capability to assign priorities to certain messages, and this flexibility needs to be built into the system. The following sections deal with three key end-to-end design issues for wireless systems: application-level adaptations, quality of service (QoS), and system security.

2.3.1 Application-Level Adaptation

A system can adapt to the variability in mobile client applications in three ways. One approach is to exploit data-type-specific lossy (i.e., involving some distortion) compression mechanisms and use data semantics to determine how information can be compressed and prioritized en route to the client.[16] A second approach is on-the-fly adaptation involving the transcoding of data into a representation that can be handled by the end application. The third approach is to push the complexity away from the mobile clients and servers into proxies, which are often used in wired networks but are not currently optimized for wireless applications.

Introduced in response to security concerns, the proxy approach is a new paradigm for distributed applications. A proxy is an intermediary that resides between the client and server—outside the client's security firewall—to filter network packets on behalf of the client. Proxies provide a convenient place to change data representations en route to the client (thereby mitigating the lossy, constrained bandwidths of wireless links), perform type-specific compressions, cache data for rapid re-access, and fetch data in anticipation of access. By supporting the adaptation to network variations in bandwidth, latency, and link error rates as well as to hardware and software variations, proxies enable client applications running on limited-capability end nodes to appear as if they were running on high-end, well-connected machines. Low-end clients (e.g., PDAs) have limited processing capabilities due to small displays and memory, relatively slow processors, and limited-capability software and run-time environments.

2.3.2 Quality of Service

Quality of service refers to traffic-dependent performance metrics—bandwidth, end-to-end latency, or likelihood of message loss—that a connection must have or can tolerate for the type of data transmitted. A network's admission-control mechanisms, which are invoked whenever a new connection is initiated, provide assurance that QoS requirements will be met; a new connection is aborted if its QoS requirements cannot be met. Attention to QoS issues is increasing because of at least two converging trends: the growing market for applications (e.g., video) that require real-time service, and the evident interest in using the Internet for a range of activities that are critical to both the public and private sectors.

2.3.2.1 Approaches to Quality of Service

Within the Internet, three categories of QoS are currently defined: guaranteed, predictive, and best-effort service. Guaranteed service is achieved if the connection conforms to a well-specified traffic specification. If the network determines that it can support this traffic, then it allows the connection to be established and guarantees that its requirements will be met. Guarantees of this type are required when the application needs tight, real-time coupling between the end points of the connection. An example of guaranteed service is the telephone system, which is designed to meet the level of perceived audio/speech quality, end-to-end switching delays, and likelihood of call blocking required for telephone calls. Certain values for these metrics are determined and the network is designed to offer the desired number of simultaneous connections. Another example of guaranteed service is that provided by ATM, a transmission protocol that handles voice, data, and video. If absolute guarantees are too expensive, then it is often sufficient to provide predictive service, indicating that the application's requirements are highly likely to be met.

The Internet as it exists today does not provide explicit QoS to different packet flows; instead it is based on a best-effort model that makes no performance guarantees. (Flows are groups of packets that share common characteristics, including the tolerated delay.) Best-effort services are appropriate for applications that do not demand real-time performance and that can adapt gracefully to the bandwidth available in the network. Best-effort service tolerates simple network components and is a good match for data transmitted in interactive bursts, interactive bulk transfer, and asynchronous bulk transfer. The common Internet data transfer applications are sensitive to losses but tolerant of latency. However, the reverse is true for emerging real-time Internet applications, which are tolerant of losses but sensitive to latency. The IETF is working to provide guaranteed services on the Internet (Peterson and Davie, 1996; Tanenbaum, 1996).

The Internet carries two broad classes of applications: delay tolerant and delay intolerant. The former applications, such as file transfer, tolerate some packet losses and delays. For these services, which are common today, the network does not reserve resources or limit the number of transfers in progress at any one time. Instead, it shares the available bandwidth among all the active applications as best it can. This is the so-called best-effort service. Delay-intolerant applications require data that is delivered with little or no delay. For these applications, different services are being developed. The components of these services consist of traffic and network descriptions, admission control procedures, resource reservation protocols, and packet scheduling mechanisms. These specifications are associated with flows. Given a service specification, the network can admit a new flow or deny access when the specifications exceed what the network can provide. The network can also police a flow to ensure that it meets the traffic specification.

Real-time Internet applications are developed on top of the real-time protocol (RTP). With RTP, a node moderates its transmission rate based on periodic reports of successfully received data at the receiver. If the sender's rate exceeds that which is reported as received, then the sending rate is reduced. Periodically the sender probes the network by attempting to increase the rate to see if a higher rate can be supported. In this way, the sender and receiver adapt to the available bandwidth without requiring any special support from the network itself. The RTP protocol is appropriate for real-time data streams, such as video and audio, that can tolerate some losses.

The real-time Internet services proposed by the IETF use the resource reservation protocol (RSVP), which enables dynamic changes in QoS and permits receivers to specify different QoS requirements.[17] The RSVP protocol is closely integrated with multicast services in which receivers determine a path through the network on which senders distribute their traffic specifications and receivers distribute their network service requirements. These sender-directed path messages and receiver-directed reservation messages are built on top of the existing multicast protocols. Senders and receivers are responsible for periodically signaling the network about their changing specifications. Once the reservations have been made, the final step is to implement them in every router on the path from sender to receiver through packet classification and scheduling. The router implementation achieves the performance specified by the network end points. The classification process maps packets on flows into their associated reservation, and scheduling drives queue management to ensure that the packets obtain their requested service.

Proponents of ATM networks view them as the foundation of integrated voice, video, and data services because they combine flexibility with performance guarantees. The ATM approach involves breaking up data into short packets of fixed size called cells, which are interspersed by time division with data from other sources and delivered over trunk networks. An ATM network can scale up to high data rates because it uses fast switching and data multiplexing based on these fixed-format cells, which contain 48 bytes of traffic combined with a 5-byte header defining the virtual circuits and paths over which the data are to be transported. The virtual circuits need to be established before data can flow. In setting up connections, the network makes resource allocation decisions and balances the traffic demands across network links, thereby separating data and control flows and enabling switches to be simple and faster.

Two types of guaranteed service are available on ATM networks: constant bit rate (CBR) and variable bit rate (VBR). The CBR service is appropriate for constant-rate data streams that demand consistency in delay. An example of such a data stream is telephone traffic that uses constant-bit-rate encodings for audio and places bounds on delivery latency. The VBR service is appropriate for traffic patterns that have a fairly sustained rate but also may feature short bursts of data at the peak transmission rate. Burst-type traffic tolerates higher delays and higher variations in delay than does constant-rate traffic. Two types of best-effort traffic classes are available on ATM networks: available bit rate, which guarantees zero losses (but makes no other guarantees) if the source follows the traffic management signals delivered by the network; and undefined bit rate, which provides no performance guarantees.

The connection orientation of ATM presents problems when dealing with network mobility. Movement between cells requires existing connections to be reestablished: Even brief transmissions invoke the full latency of connection setup. In addition, ATM is not appropriate for lossy links because there is no agreed-upon mechanism for error recovery or retransmission at the link layer. Considerable controversy exists as to whether ATM will be used throughout a system or only at the link or subnetwork level. Full ATM connectivity, from one end of a system to the other, is required to take advantage of the service guarantees.[18] Many believe that it will be necessary to run TCP/IP over ATM to ensure interoperability across heterogeneous subnetworks. Debate continues over how to interface ATM's performance guarantees with the emerging Internet capabilities for predictive service.

Wireless communications introduce additional QoS issues. The QoS guarantees for expected loss rates, latencies, and bandwidths were developed based on the assumption that switched, fiber-optic wired networks would be used. Such networks feature low link-error rates, easily predicted link bandwidths, and QoS parameters that are largely determined by how the queues are managed within the switches. As a result, losses are due almost entirely to congestion-related queue overflows. Wireless links, on the other hand, have high bit-error rates, high latencies due to link layer retransmissions, and unpredictable link bandwidths.[19] Furthermore, the quality of a wireless link varies over time, and connections can be lost completely. Two wireless end nodes sharing the same link can experience vastly different link bandwidths depending on their relative proximity to the base station, location in a radio fade, or loss of receiver synchronization in a multipath environment. Link quality can also be degraded by interference from a nearby transmitter. In addition, hidden terminals cause time-consuming back-off (i.e., waiting before resending) that further degrades network performance.

Because link quality varies on small time scales, it is difficult to improve a wireless link through agile error coding or increased transmit power. Moreover, attempts to improve a link for one user can adversely affect others. Guarantees are elusive in this complex environment.[20] The ATM end-to-end QoS model is difficult to implement if the limiting link is wireless. In general, however, approaches such as adaptive spreading codes, FEC, and transmit-power control can be used at the media access control and link layers to improve higher protocol layers in a wireless environment (Acampora, 1996).

2.3.2.2 *Transport-Layer Issues*

The most widely used reliable transport protocol is TCP, a connection-oriented protocol that combines congestion control with "sliding-window" flow control at the sender and cumulative acknowledgments from the receiver. As each segment is received in sequence, the receiver generates an acknowledgment indicating the number of bytes received. In the current generation of TCP, congestion is controlled by the sender, which maintains a variable called the congestion window that regulates how much data the sender can have in flight across the network at any one time. The sender adjusts the congestion window in response to perceived network conditions.

When a TCP connection first starts, or after a major congestion event, the congestion window is set to one packet, which means the sender cannot send a second packet until it receives an acknowledgement of the first. The sender then adjusts the congestion window by doubling it each round-trip across the net. This part of the algorithm is called slow-start. Once a certain threshold is crossed, the congestion-avoidance phase is triggered and the window size grows in increments of a single packet for each round-trip. The sender uses these mechanisms to probe the network to discover how much data can be in flight.

A lost packet creates a gap in the sequence number of data arriving at the receiver. When this occurs, the receiver generates a duplicate acknowledgment for the last segment received in order. When a threshold number of duplicate acknowledgments is received, the sender retransmits the lost segment and halves the congestion window; this part of the algorithm is known as fast retransmission and recovery. A more serious congestion event can cause the loss of so many packets that no duplicate acknowledgments are generated by the receiver. The sender detects and corrects this situation using a timer. The TCP protocol sets time-outs as a function of the mean and standard deviation of the round-trip time. If no acknowledgment is received within this interval, then the sender retransmits the first unacknowledged segment, sets its window to one packet, and reenters the slow-start phase. This event causes a major reduction in throughput until the window opens and also can cause a silent period until the timer expires.

Fast retransmission works well in many circumstances today. However, specific issues arise in wireless systems. When a packet is lost or damaged because of bit errors on the wireless link, this loss is detected and corrected by the fast-recovery algorithm. However, fast retransmission reduces the window size as a side effect, thus keeping throughput low. These problems can be mitigated through a TCP-aware link layer (Balakrishnan et al., 1996), in which the base station triggers local retransmissions of lost segments. By intercepting the duplicate acknowledgments, the base station shields the sender from the effects of local losses that would have the effect of shrinking the congestion window and reducing throughput. More seriously, burst errors on the wireless link can cause the loss of several packets, which will trigger the slow-start algorithm even though there is no congestion.

Asymmetric connections, in which the bandwidth in one direction far exceeds that available for the opposite path, can present problems for transport-layer connections because the effective bandwidth on the forward path is limited by the amount of acknowledgment traffic that can be sent along the reverse path. An example of asymmetry is direct-broadcast satellite, which sends data to the user at several hundred kilobits per second but uses substantially smaller-bandwidth technologies (such as conventional telephone or wide-area wireless) for the return path at tens of kilobits per second. Asymmetries also arise because of the nature of data traffic patterns. For example, Web access involves much more data transmitted from servers to users than in the opposite direction. Yet poor performance on the acknowledgment path moderates the performance on even a high-bandwidth forward path. One solution to is to compress the acknowledgment packets; another is to delay acknowledgments so that each one acknowledges an expanded range of received data.

Asymmetries can undermine reliability in the application of TCP to wireless links. The TCP protocol can adapt the duration of its time-outs as long as the round-trip time estimate is not highly variable. Asymmetries in the connection bandwidth, coupled with differential loss rates and congestion effects on the forward and reverse paths, increase the variability in estimated round-trip time. This means that when losses do occur, the retransmission time-outs can become very large, significantly degrading a connection if losses occur often.

2.3.3 Security

Wireless communication systems are inherently less private than are wired systems because the radio link can be intercepted without any physical tap, undetected by the transmitter and receiver. Wireless networks are therefore especially vulnerable to eavesdropping, usage fraud, and activity monitoring, threats that will grow as wireless banking and other commercial services become available. In addition, both wired and wireless networks need to be designed to maintain the integrity of data and systems and assure the appropriate availability of services. Thus, security is an important issue for both commercial and military applications. For purposes of this discussion, which considers key aspects of the information security challenge but is not comprehensive, the issues can be divided into three categories: network security, radio link security, and hardware security.[21]

Network security encompasses end-to-end encryption and measures to prevent fraudulent network access and monitoring. One user-oriented framework distinguishes several levels of end-to-end encryption (Garg and Wilkes, 1996). Level 0 has no encryption, meaning that anyone with a scanner and knowledge of the communication link design can intercept a transmission. Analog cellular telephones offer this level of security, which has been a problem and has motivated security upgrades in the digital cellular standards.[22] Level 1 provides low-level security such that individual conversations might take a year or more to decrypt. This level is probably secure enough for commercial telephony applications, provided that an equivalent effort would be needed to decrypt subsequent conversations ("perfect forward secrecy"). Level 2 provides increased (perhaps by a factor of 10) security for sensitive information related to electronic commerce, mergers and acquisitions, and contract negotiations. Level 3 provides the most stringent level of security, meeting government and military communications requirements as defined by the appropriate agencies.

Radio link security prevents the interception of radio signals, ensuring the privacy of user location information and, for military applications, AJ and low-probability-of-detection-and-interference (LPD/I) capabilities. Link security was primarily a military concern before commercial wireless communications became prevalent. Military systems are designed to avert the detection of radio signals, jamming of communication links, and interception and decoding of messages. Many military radios are based on spread-spectrum technology, which provides both AJ and LPD/I capabilities. However, because knowledge of the spread-spectrum code would enable an adversary to intercept a spread signal, encryption is usually applied as well to prevent signal interception and message recovery. Many military techniques for reducing interception and detection are classified.

For commercial systems the primary link-security issue is privacy, which is not typically assured. Conversations on analog cellular telephones are accessible to anyone with an FM scanner, as demonstrated by recent publication of communications involving public figures. Moreover, the location of a cellular user can be determined by triangulating the signal from two or more base stations, a feature that has been exploited successfully by law enforcement authorities. It is difficult to prevent the interception of commercial radio signals, not only because communications protocols are publicized in patents and standards but also because most communications devices have a "maintenance" mode for monitoring calls (a capability intended for testing purposes that could also be used to eavesdrop).

It is unlikely that commercial devices will ever require a level of security equivalent to military systems and may not even provide the "hooks" enabling the addition of LPD/I capabilities. Similarly, although the growing use of wireless systems and growing dependence on networked communications have heightened concerns about the possible denial of service in commercial contexts, there is probably greater tolerance for private-service outages than for jamming in a military situation at this time.

Hardware security also has different implications for commercial and military applications, although encryption keys typically need to be protected in both contexts.[23] Commercial systems require sufficient security to prevent the fraudulent use of information in the event of theft or loss, and user databases need to be secured against unauthorized access. The military has similar requirements but at a much higher security level. It also has additional requirements: Military devices need to be protected so that opening them will not reveal any of the specialized hardware or software technology.

2.4 HARDWARE ISSUES

Among the hardware issues that are critical to third-generation wireless systems, radio stands out as being central to the military mission. The radio receiver consists of an antenna, RF amplifier, mixer, filters, demodulator, and decoder (see Figure 2-3). Radio signals are received by the antenna, amplified, passed through the mixer and filters, demodulated, and decoded. Transmitters have similar architectures but the op-

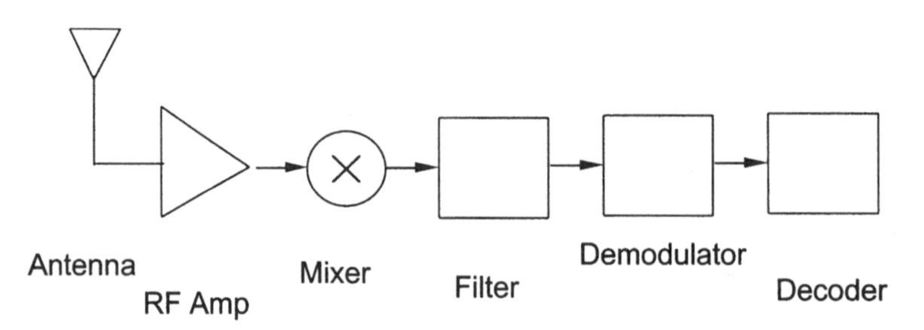

FIGURE 2-3 A radio receiver has six basic components. In transmit mode the operations proceed in the reverse order.

erations are performed in reverse order: The data are encoded, modulated, passed through the filters and mixer, amplified, and transmitted through the antenna.

Section 2.4.1 reviews antenna technology, which has become increasingly sophisticated with the addition of adaptive capabilities. Section 2.4.2 discusses other radio components, emphasizing the transition from analog to digital technology and from single-purpose to multipurpose systems. Traditional radios were designed for a single air interface (i.e., one modulation type occupying a particular bandwidth). Given the proliferation of standards and the need for compatibility with older equipment, the general trend in radio design is to build flexible systems that can handle multiple air interfaces. Section 2.4.3 discusses portable terminals. In modern mobile devices, the radio system is integrated with sophisticated user interfaces and computing capabilities in lightweight, modular packages. The design of portable terminals relies on advanced microprocessors, displays, user interface devices, power sources, and software.

2.4.1 Antennas

An antenna serves as the interface, or transducer, between the electronic circuitry of a transmitter or receiver and the medium through which radio waves travel. Classical antenna designs include simple stub or "whip" antennas such as those found on cellular telephones, as well as massive, parallel panels that are aligned in phase to provide flexible electronic steering (examples include the phased-array radars used on some warships). While in transit between the transmitter and receiver, the RF signals are subject to a variety of distortions (see Section 2.1.1.3). In addition, they create interference for other communications and provide opportunities for interception. To limit interception and interference and also to conserve power, antennas can be designed so that the RF energy radiates in only a particular direction, providing gain along the intended direction and attenuation in undesired directions.

Various antenna structures have been developed to direct electromagnetic signals. Receiving antennas also have directional properties: The most common examples are rooftop television antennas that point in the direction of the local television transmitter and satellite dishes that point at the orbiting satellite. Directional antennas in cellular-system base stations focus power in a particular direction, thereby minimizing the required transmitter power and significantly reducing the amount of interference. Directional antennas need to be positioned carefully. Positioning is not difficult in local television broadcasting because both the transmitter and receiver are stationary and a fixed, narrow beam works well. In cellular and personal communications systems, however, the transmitter and receiver locations are mobile. Therefore, most directional antennas in mobile communications have a fairly large beam width (60 to 120 degrees). Although narrower beams would enable the use of low-power transmitters and reduce interference, many such beams would be required to cover even a small service area, and mobile users would constantly be moving from one beam to another.

User mobility is a key motivation behind the development of steerable antennas—so-called smart or adaptive-array antennas—that can change the shape and direction of their transmission beams depending on user location. A steerable transmitting antenna controls the phases of the electromagnetic signals generated at each of its numerous elements, thereby changing the physical locations at which the wave-like signals add constructively (to create a beam) or destructively (to create a null). Using feedback control, an antenna beam can follow the movement of a mobile unit. Spot beams can be created by both the transmitting and receiving antennas. These technologies increase system capacity, reduce transmitter power requirements and interference, and dramatically reduce the likelihood of unwanted signal interception. Future mobile systems will pinpoint the relative positions of the transmitter and receiver with even greater accuracy than is currently possible, making sophisticated location-based services feasible.

The physical size of the elements in an antenna is related to the wavelength of operation, which is inversely proportional to the transmission frequency. Thus, higher operating frequencies mean shorter wavelengths, smaller antenna features, more elements per antenna, and the possibility of more complicated and precise beam patterns. Adaptive antennas are already used in military operations, particularly at frequencies above 20 GHz, to accommodate very wideband signals used for communications, tracking, or guidance. Such antennas are composed of many elements and are fully capable of electronic beam forming and steering. Each element is controlled electrically through changes in the properties of dielectric materials; the antenna does not change physically and therefore needs no moving parts.

Older communications systems have a number of shortcomings. Some have multiband capabilities but are expensive and bulky. Commercial applications for adaptive antennas are limited to relatively low-cost, single-band units with limited flexibility in beam pattern. Moreover, virtually all existing adaptive antennas for mobile radio applications are designed for use at base stations rather than mobile units. The key technical challenges in the design of adaptive antennas for military applications are to reduce the size and cost of the RF and signal-processing technology and achieve additional gain in a handset by designing three-dimensional instead of planar designs. The commercial sector is likely to need such designs in

TABLE 2-2 Use of Digital Components in Commercial Communications Products

Product	RF[a] Amplifier	Mixer	Filter	Demodulator	Decoder
Car radio with equalizer	Analog	Analog	Analog	Analog	Analog
DirecTV receiver	Analog	Analog	Analog	Digital	Digital
Dual-mode cell phone	Analog	Analog	Digital	Digital	Digital
PC telephone modem	Analog	Digital	Digital	Digital	Digital
FDDI[b] modem	Digital	n/a	n/a	Digital	Digital

[a]Radio frequency.
[b]Fiber-distributed data interface.

the future as new spectrum is allocated at higher frequencies (above 2 GHz) and multiband radios become available. For the time being, the DOD may need to fund its own R&D in this area.

2.4.2 Other Radio Components

The evolution of digital technology is transforming radios. Other than antennas, all the components of the radio system—RF amplifier, mixer, filter, demodulator, and decoder—are amenable to either analog or digital implementation. Many commercial radios and other communications products already use programmable digital modules (see Table 2-2).

There are many advantages to replacing analog hardware with programmable digital technology, although trade-offs are involved. As noted above, digital technology offers inherent security advantages. Another benefit is time to market: As with PCs, product development time can be reduced because changes and improvements can be implemented through software. Digital technologies also make it easier to achieve temperature stability and reliability and to manufacture, support, and test equipment. Digital radios can be designed for performance peaks, whereas analog radios de-optimize performance because the filters are detuned to make the system easier to manufacture and tolerant of component variability. Finally, digital components can reduce costs by providing increased functionality per unit and reducing the need for multiple types of radios.

The design of wideband (i.e., multiband) digital radios has been enabled by rapid advances in microelectronics, including DSPs, A/D converters, ASICs, and field-programmable gate arrays (FPGAs). In new radio architectures, referred to variously as software-defined radio, programmable radio, or simply software radio, analog functions such as tuning, filtering, demodulating, and decoding are replaced with software directing the digital equivalents. The mixers and filters can process multiple modulations spanning multiple bandwidths; the demodulation and decoding processes are programmed; and modulation and coding are usually performed using DSP chips.

The wideband A/D conversion of the software radio enables the implementation in handsets of direct frequency conversion (i.e., eliminating the typical intermediate steps between the baseband and transmit frequencies, thereby reducing noise and the need for filtering). This design is not yet appropriate for commercial systems, because it is not feasible at this point to service a large number of subscribers using such receivers simultaneously. In the handset, the RF amplifier is required to obtain a reasonable noise figure and input intercept. The anti-alias filter selects which of the multiple subbands to digitize. The wideband (multiband) digitizer converts all RF signals into a digital representation. The processor uses software to implement all legacy and future radio systems. The processor is capable of implementing multiple simultaneous radios, much like a PC can run multiple applications simultaneously.

The use of digital radio hardware still presents challenges and requires trade-offs that may not be readily apparent. A digitally implemented radio needs to be at least as good as the analog radio it replaces in terms of QoS parameters such as reception sensitivity or power. This challenge is being met: Coding and decoding improvements, driven by DSP advances, are making digital radio systems not only equal to analog systems but also better. However, four limiting technologies need to be developed further if wideband (i.e., multiband) software radios are to become a practical reality: advanced A/D converters, DSP chips, filters, and RF amplifier components.

2.4.2.1 *Analog-to-Digital Converters*

The key enabling component, and the most complex and misunderstood element of wideband software radios, is the A/D converter. Most A/D converters are characterized by maximum clock rate and number of output bits, digital metrics similar to those used to characterize microprocessors or memory devices. However, because signal quality is key, A/D converters are better characterized using analog characteristics such as SNR, signal range free from spurious noise, and usable bandwidth (known as Nyquist bandwidth). The use of these metrics helps ensure that the critical A/D transition can be accomplished with minimum degradation in signal quality.

An A/D converter can be implemented in many different architectures; the three significant modern architectures are known as flash (or parallel), subranging, and sigma-delta. The choice involves trade-offs between the accuracy and conversion rates. The flash A/D uses the most hardware and power but also operates at the fastest sampling rates and produces the largest usable Nyquist bandwidth. These converters are generally inadequate because of a lack of dynamic range. Subranging A/D converters are slower but offer both the bandwidth and dynamic range needed for software radios. Sigma-delta A/D converters generally have very low sample rates and are appropriate for narrowband applications. It is not clear at this stage which technology can be improved most readily to provide the requisite low-power converters with high dynamic range that process bits as rapidly as possible. A demonstration of ultrafast A/D converters was planned as part of the DOD-funded Millennium program. The commercial sector is also performing R&D in this area and is likely to produce advances that would be appropriate for military applications.

2.4.2.2 *Digital Signal Processors*

For the past few decades semiconductor manufacturing has followed Moore's Law, which predicts that the number of devices on an IC will double every 18 months. This trend is directly related to the steady reductions in the minimum feature size, or linewidth, that can be manufactured in large volumes. The Semiconductor Industry Association road map calls for the production of devices with more than a billion transistors by the year 2010. Such densities could enable entire systems to be built on a few or even a single chip. Indeed, software radio architectures may be implemented increasingly in small numbers of ICs (see Figure 2-4).

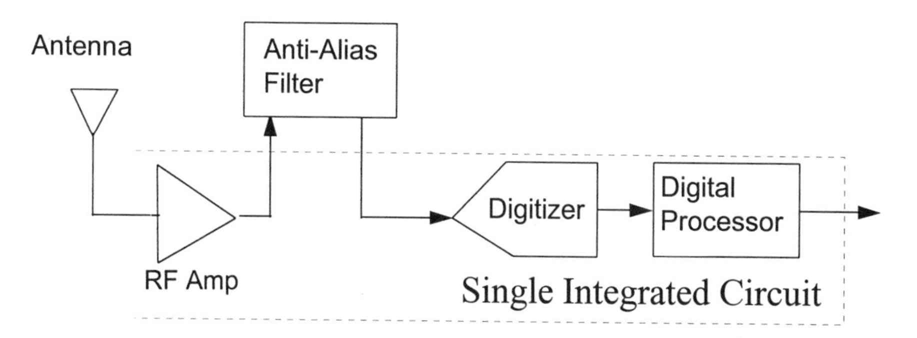

FIGURE 2-4 Future wideband (i.e., multiband) software radios may implement some functions, such as analog-to-digital conversion and signal processing, in single integrated circuits.

The several general-purpose, programmable DSPs now on the market are primarily a customized segment of the IC industry. Although DSP speed is improving every year, single-chip performance is still very limited for software radio applications. High speed can be achieved with large arrays of DSPs but the size, weight, power, and cost of this design are not attractive for small or handheld radio applications. Because filters are critical to the performance of a software radio, gains could be achieved through the use of integrated FIR-filtering ICs. These devices, developed very recently, could perform the processing function at a small fraction of the complexity and cost of a programmable DSP.[24] Regardless, the rapid commercial advances in signal processing technology are likely to produce chips that meet military needs.

2.4.2.3 Filters

Filters influence not only a radio's signal-processing speed but also its sensitivity, dynamic range, and capability to avert co-site interference. Their importance is reflected in their physical presence: Filters constitute 25 percent of the volume of a typical software radio, in part because several different filters are needed (i.e., for receive preselectors, amplifier output, local oscillators, and mixers). Improvements in frequency-tuning range and selectivity as well as miniaturization would be helpful, especially for application in handheld devices. The commercial sector continues to rely on older technology (e.g., mechanical filters are used in cellular telephone systems) whereas the military has unique needs to reduce co-site interference, both within software radios and across multisystem platforms, and cover wide frequency ranges. Existing radios that span wide frequency ranges require combined filters made of new materials that have remarkably flexible and adaptive electrical properties, far beyond older static inductors and capacitors. The new materials and modern filter fabrication techniques will lead to new and smaller implementations of wideband filtering based on the fundamentals of transmission-line techniques. Thus, filters may merit a significant military R&D investment.

2.4.2.4 Radio Frequency Amplifiers

The commercial sector is designing ultralinear amplifiers that will process many signals from multiple transmitters and add them coherently to achieve good fidelity. These designs will improve power efficiency and consume less space than traditional amplifiers. However, the commercial sector is unlikely to produce multiband amplifiers, which will be very expensive, anytime soon. Alternative materials might offer advantages in the design of future military systems and could be a topic for DOD-funded research.

The most important recent development in RF technology is the re-emergence of semiconductors (i.e., silicon) as an alternative to the semi-insulator materials (e.g., gallium arsenide) traditionally used for RF device manufacturing. Semiconductors offer two advantages. First, they form a natural ground plane on an IC such that microwave devices can be fabricated much closer together, resulting in smaller chips and enabling the design of circuits that cost less and support higher frequencies and performance than do conventional circuits. Second, semiconductors cost less than semi-insulators because they are produced in higher volumes (they are also used in the most advanced CMOS microprocessors and random-access memory [RAM] chips).

As a consequence, silicon is now being used in moderate-performance RF front ends for cellular and personal communication systems. Further, a new technology involving the implantation of germanium atoms in silicon to create heterojunction bipolar transistors promises an extremely low cost, silicon-based approach for RF (30 MHz to 2 GHz) and microwave (2 GHz to 40 GHz and above) analog front ends and power amplifiers.

2.4.3 Portable Terminal Design

The small size and portability of wireless communicators provide obvious benefits for users but also introduce challenges for system designers because they limit display, processing, power, and storage capabilities. The following subsections review the limitations and the new technologies designed to overcome them. The commercial sector is making rapid advances in all these areas that the DOD can exploit to good advantage.

2.4.3.1 Displays, User Interfaces, and Input Devices

Small, highly portable devices contain relatively low quality displays. There are three reasons for this. First, portable devices have limited physical space and power available for the display. Second, display pixels cannot be smaller than the resolving limit of the human eye, meaning that the number of pixels in a given display (i.e., the resolution) is limited. In addition, bright colors can be produced only if there is sufficient power for backlights and display elements; otherwise the display is dim and monochrome. For these reasons the user interface of a portable device needs to be designed for monochrome presentations in a very small screen area—a significant impediment to the display of video or high-quality images. Nevertheless, significant market forces are fueling a trend toward ubiquitous information displays, and commercial displays offering high resolution, full color, and reduced power requirements are likely to be developed.

Because portable devices lack the space for standard keyboards, icon-based interfaces and pen-based input have been considered as alternatives. In some devices the keyboard is replaced with a small number of function-specific buttons. These devices still support functionally specific virtual keyboards, which are displayed on a touch screen and can be operated by applying pressure to the keys with a stylus. The pen-based devices either support handwriting recognition or simply capture pen strokes ("digital ink").

Ideally, mobile communication devices will be able to send images from remote sites. This capability will be enabled by charge-coupled devices (CCDs)[25] and CMOS camera chips. These chips are already used in commercial camcorders and have become inexpensive and widespread as a result. Highly integrated cameras have been declining in price, and such a camera is integrated into at least one state-of-the-art Japanese PDA.

2.4.3.2 *Processors*

The successful development of low-power devices with long battery life has placed limits on the raw performance of embedded processors because processing speed and clock cycle directly influence power consumption. New metrics are therefore required to measure the performance of processors for portable applications: millions of instructions per second (MIPS) per watt, a measure of the impact on battery life and heat dissipation in highly integrated systems; MIPS per square millimeter, a measure of the silicon manufacturing costs of the processor; and bytes per task, a measure of the amount of memory that devices need to incorporate to perform signal-processing functions. Because consumers are demanding highly integrated yet portable computing devices, the commercial sector is performing R&D with the aim of increasing processor capabilities while also reducing power requirements.

2.4.3.3 *Batteries*

The commercial sector has made tremendous strides in battery technology in recent years because it plays a role in many technologies, ranging from surgical implants to electric cars. Nickel cadmium (NiCd) batteries are the most widely used rechargeable batteries, found in many consumer electronic devices. Most laptop computers now use nickel metal hydride batteries, which have slightly better energy storage per weight and substantially improved energy storage per volume. Lithium ion (LiIon) batteries are used in some new portable products, such as small cellular telephones. The energy-storage capacity of LiIon batteries is more than twice that of NiCd technology by both weight and volume. Lithium polymer (LiPoly) batteries are about 10 percent more efficient than are LiIon batteries. Both LiIon and LiPoly batteries use solid electrolytes, making it possible to form the battery into arbitrary shapes, a significant improvement over other battery technologies.

2.4.3.4 Storage

The disk-drive capacity of information processing devices continues to increase while physical size shrinks, but the 2.5-inch disk widely used in notebook and laptop computers is still too large for handheld devices. In PDAs the disk is replaced by RAM in the form of battery-backed-up static RAM and flash RAM on personal computer multiple component interface access (PCMCIA, or just PC) cards, which can cost 30 times more than disk storage per megabyte. Commercial R&D in this area is producing steady, impressive advances that are likely to meet military needs.

2.5 SUMMARY

The design of wireless communications systems presents countless challenges. Some solutions are available and many more are on the horizon. Although the review presented in this chapter is general in nature, consideration of this information in the context of DOD's needs suggests a number of areas deserving careful attention in the design of future military systems.

Specifically, network architecture is a fundamental issue that defines all other aspects of the system design. The basic choice is between a peer-to-peer and base-station-oriented design, but there are also other questions related to how infrastructure elements are connected and the nature of communications with other networks. The commercial and defense sectors have differed in their choice of network architectures in the past and continue to have some different needs and concerns. The selection of an optimal military network design could be assisted by simulation and modeling. However, current tools are inadequate to the task of modeling an untethered communications system that uses wideband signals and advanced components such as software radios.

The DOD also has unique needs for interoperability and security of communications systems, although commercial concerns about system integrity and service availability are growing. The evolution of software radios will enable interoperability among advanced and legacy systems, but this technology presents co-site interference problems that will require new solutions. Similarly, the available AJ and LPD/I technologies will need to be complemented with security advances that accommodate global, heterogeneous communications systems and multiple security levels. The emergence of wideband, programmable radios for military applications will also depend on advances in hardware components such as antennas, which need to be designed for mobile units, and filters, which need to be miniaturized and designed for wideband applications.

These issues are examined further in Chapter 3, which explores the opportunities for synergy between the commercial and military sectors in the development of advanced wireless communications systems.

NOTES

1. These error rates are associated with the link layer of the OSI model and are commonly accepted as tolerable for these applications. At higher levels of the OSI model the use of error-correction protocols can improve the rates.

2. These definitions apply only to unmodulated waveforms. Modulation changes the phase and frequency with time (see Section 2.1.3) such that the definitions are no longer accurate.

3. There are numerous path loss models that conform to a variety of propagation mechanisms, including free space, reflection, diffraction, scattering, or some combination of these (Rappaport, 1996).

4. The relationship is $L = Pr/Pt = K/f^2d^n$, where Pr is received power, Pt is transmitter power, f is the center frequency of the transmitted signal, and K is a constant that depends on the average path loss at a reference distance d_0 from the transmitter (d_0 is the far field of the antenna, typically 1 m for indoor environments and 0.1–1 km for outdoor environments). The exponent n is the path loss exponent.

5. This analysis is based on the assumption that the channel is changing slowly enough to allow for adaptation, and that the channel fading can be estimated accurately at the receiver and this information fed back to the transmitter with minimal delay.

6. A RAKE receiver produces a coherent sum of individual multipath components of the received signal. The components can be weighted based on their signal strength to maximize the SNR of the RAKE output. The sum provides an estimate of the transmit signal. A RAKE receiver is essentially another form of diversity because the spreading code induces a time diversity on the transmitted signal such that independent multipath components can be resolved.

7. If multiple systems share the same bandwidth without any channel access coordination and are not interoperable, then some technique is still needed to enable efficient operations. Etiquette rules permit incompatible systems to coexist when using the same bandwidth (whereas interoperability requires standardization—agreement on all waveforms and protocols before systems are built and deployed). The Wireless Information Network Forum, an industry group, has

defined etiquette rules for the unlicensed personal communications bands and has taken the same basic approach for the 60-GHz spectrum allocation (Steer, 1994). The key elements of etiquette rules are (1) listen before transmitting to ensure that the transmitter is the only user of the spectrum, thereby minimizing the possibility of interfering with other spectrum users; (2) limit transmission time to allow others to use the spectrum in a fair manner; and (3) limit transmitter power so as not to interfere with users in a nearby spectrum.

8. For many networks, including voice-oriented cellular networks, the number of transmitters active at any one time is much smaller than the total number of possible transmitters that need be recognized by the hub station. The unpredictable and dynamic nature of the set of active transmitters clearly precludes the fixed assignment of separate channels to each transmitter.

9. Simultaneous detection of multiple users is not currently possible because of the increased complexity required in the receiver. Multiuser detection schemes also require low BERs because bits that are incorrectly detected are subtracted from the signals of other users, possibly causing those signals to be decoded in error as well.

10. These analyses are based on simplifying assumptions about the hardware and communications environment; many of these assumptions would break down in a real operating environment. Moreover, it is not known which technique has a higher spectral efficiency in flat or frequency-selective fading, particularly when countermeasures to fading are used.

11. Techniques are available to avert the delay. For example, a certain number of packet slots can be allocated for unreserved transmissions using a contention scheme. The successful sending of a packet in this slot is taken as a request for a reserved slot (or two) in the next round-trip. As long as the slots are used this reservation continues to be available, and as long as there is capacity the reservation is allowed to grow. But the reservation is abandoned as soon as the sender does not use the slot. This approach involves no delay (except for contention failures on the first packet), poses contention issues only for the first packet in a burst, and matches the natural behavior of the TCP slow-start phase (which is described later in this chapter). However, for applications that alternate a short message in each direction (e.g., transaction processing) the procedure still produces latency equal to one round-trip for each message, and, assuming fixed-length slots and a perfect fit between the data to be transmitted and a slot, has a fundamental throughput limit of 33 percent. If the transmission is smaller than a slot, then the throughput will be even lower (and lower still in many realistic applications with short transaction times). If the information to be transmitted is less than or even comparable to the amount of information required to set up the DAMA resources, then efficiency will be compromised.

12. Assuming that a collision results in the loss of two packets, the maximum throughput in an ALOHA channel is about 18 percent of the peak data rate if the probability of a collision is to be reduced to a level acceptable to the user. Various modifications of ALOHA channels, such as slotted ALOHA or CSMA/CD, can increase efficiency, but they also impose restrictions on data transmission.

13. The latency of Mobile IP is typically much less than a second—the time it

takes for one round-trip between the foreign agent and home agent, or perhaps two round-trips counting the time for message receipt verification.

14. Route optimization is an enhancement to the base specification for Mobile IP and has not to date reached an equivalent level of standardization within the IETF. Mobile IP is a proposed standard (RFC 2002–2006), whereas route optimization has yet to be standardized or shown to be interoperable in multiple implementations.

15. A typical simulator accepts as input a description of network topology, protocols, workload, and control parameters. The output includes a variety of statistics, such as the number of packets sent by each source of data, the queuing delay at each queuing point, and the number of dropped and retransmitted packets. Visualization packages have been developed to allow the simulator's dynamic execution history to be made visible to the network designer. The simulators are designed in such a way that they can modified easily by users.

16. Recent research has investigated the adaptation of compression algorithms to a channel of varying quality (i.e., a channel with fading or varying noise or interference levels). Such adaptation can reduce distortion significantly. This design is based on the idea that, because the transmission rate is constant, this rate needs to be divided between the compression algorithm and the channel code. The optimal way to divide the transmission rate and minimize distortion is the following: On a channel with high SNR, no channel coding is needed, and all the rate is allocated to the compression scheme; as the channel quality degrades, more of the rate is allocated to the channel coding to remove most of the effects of channel errors. However, joint compression and channel coding creates some problems. First, this approach requires that the compression algorithms, which typically sit at the application layer, have access to information about the link layer, which means that the layer separation of the open-systems interconnection model breaks down. Second, the design can become very complicated. It is often easier to design the compression algorithms and the channel coding independently and then "glue" them together (the compression and coding communities prefer this approach because they have developed separate languages and perspectives, which make it difficult for them to work together). Some future cellular systems will implement a crude form of this joint design using "vocoders" (compression schemes for voice) that operate at multiple rates. If the channel has a high SNR, then the higher-rate vocoder (which performs poorly at low SNRs) is used, and the vocoder rate is decreased as the channel quality decreases.

17. The Internet community is carrying forward two proposals for real-time service: Guaranteed service provides per-flow hard guarantees (i.e., no statistical aggregation or probabilistic bounds), whereas controlled-load service provides a probabilistic bound based on aggregation of a number of real-time flows into one scheduling class. Although guaranteed service provides a delay bound that is computed in advance, controlled load provides a bound that is stable but not explicitly computed. The application must adapt to the service it receives. Both are set up using RSVP. The soft and hard states differ in terms of what happens when a route fails. In ATM the connection is cleared and no traffic is delivered until a new connection is established. In Internet/RSVP the packets start flowing once the routing tables have found a new route, but only with default QoS until

RSVP reestablishes the state. In both cases the new request may fail if there is not enough capacity after the failure.

18. The Wireless ATM Working Group of the ATM Forum (an industry group) is addressing the problems of end user mobility. This effort may be the only avenue for extending ATM to the end user.

19. Latencies in the wireless channel are not only high but also variable over time because of fluctuations in retransmission. Forward error correction can mitigate this problem somewhat but imposes a penalty even when the channel quality is good.

20. Because of the error characteristics of wireless links, some of the QoS issues need to be addressed locally at the link layer rather than from an end-to-end perspective. The DARPA PRNet had a strategy of accomplishing enough at the link level that TCP could handle the remaining reliability issues. However, this approach requires interaction between the link layer and higher layers (e.g., if the link layer needs to implement a stronger channel code, then its transmission rate may be reduced or its delay increased). In addition, the wireless channel may be so degraded that little can be done at the link level to improve matters. There needs to be a way to cope with this situation through higher-layer protocols.

21. Software security is another category but it is not unique to wireless communications and therefore is not addressed here.

22. Some security concerns are being alleviated in the transition from analog to digital systems, which offer an inherent advantage because the meaning of a pattern of 1s and 0s cannot be casually discerned.

23. For example, systems based on the GSM standard keep the key in a separate smart card, not in the telephone.

24. For example, most contemporary software radios use commercial filters by Graychip, Inc., or Harris Corp. for highly programmable channel access to FDMA, TDMA, and CDMA systems with the low size, weight, and power of ASICs.

25. A CCD detector turns light into an electric charge, which is then transformed into the binary code recognized by computers. Some commercial cameras use this technology, but they remain expensive.

3

Commercial-Defense Synergy in Wireless Communications

Wireless communications technology development is a complex process that includes interactions between the commercial and military sectors. An understanding of these interactions, including the opportunities for and barriers to synergy, is crucial to an evaluation of the potential for expanded military use of commercial products. Building on the historical and technical foundation provided earlier in this report, this chapter identifies broader organizational and R&D issues that need to be addressed to ensure that the DOD fields affordable, state-of-the-art untethered communications systems that meet future military needs.

Wireless technologies are often transferred among government, industry, and academia. Such interactions take place through multiple mechanisms, sometimes in a continuing cycle from the commercial to the defense sector and back again (see Box 3-1). The synergy can evolve either during the initial research or after technologies are developed. For example, the DOD's funding of basic academic research on wireless technologies and networking (currently through the DARPA GloMo program) creates an active technology base for use in both military and commercial industries. Similarly, there is overlap within companies that have both commercial and defense divisions. Most large corporations also support academic research to gain access to important new concepts.

This chapter examines how this synergistic process might be leveraged to meet future military needs in untethered communications. Section 3.1 provides a brief overview of military use of commercial wireless products. Section 3.2 identifies the motivations and opportunities for

BOX 3-1
Handie-Talkies Serve Both Military and Commercial Needs

In 1940 Motorola developed the first handheld two-way radio, the Handie-Talkie, a 2.3-kg AM unit with a range of 1.6 to 4.8 km. Within three weeks of U.S. entry into World War II, Handie-Talkie production exceeded 50 units a day; by 1945 more than 130,000 units had been built. In 1942 Motorola's design for the world's first portable FM two-way radio, the SCR-300 backpack unit, won a competition to replace an older Army Signal Corps radio, the "walkie-talkie." The SCR-300 weighed almost 16 kg, had an average range of 16 to 32 km, and could be tuned to various frequencies in the 40–48 MHz band. Motorola police radios were used in the Army's first radio relay system for behind-the-lines communications and its first radio teletype hookup. After the war, Motorola introduced the first commercially available portable radiophones, the Handie-Talkie radio line. A fully transistorized, VHF pocket transmitter version was developed in 1960. A fully transistorized, portable two-way radio was developed in 1962; its weight of approximately 1 kg was reduced by almost half in 1969. These devices have evolved into Motorola's current line of cellular telephones. Component technologies from commercial communications equipment are now designed into future generations of military equipment, thus furthering the ongoing cycle of commercial-defense synergy.

commercial-defense synergy in the development of wireless technology. Section 3.3 outlines the barriers to synergy posed by mismatches between commercial capabilities and military needs and operating requirements. Section 3.4 examines three broad issues that need to be addressed in the design of future wireless systems for defense applications. Section 3.5 reviews the relevant defense technology policy issues.

3.1 OVERVIEW

Myriad wireless technologies have originated within the government. Satellite programs initiated by the federal government in the early 1960s produced technologies that were quickly adopted for commercial use, starting with INTELSAT in 1965 in the United States and other countries in the 1970s. Another important government-initiated technology was packet switching, developed by DARPA (then known as ARPA) in the late 1960s. This advance led to commercial and military packet-switched systems worldwide as well as to the Internet. The government also led the work on advanced coding techniques (for recovering data from deep-space probes), spread-spectrum techniques, signal and data encryption, and more recently on-board digital processing. All of these technologies have been adopted by commercial enterprises.[1]

Conversely, the U.S. military uses many commercial communications products. The military uses a variety of commercial systems, including satellites developed in the mid-1970s to transfer weather data to computer processing centers and disseminate the processed data; commercial satellites and land-based services to transport military-encrypted communications links; VSAT networks operating over commercial satellites to disseminate logistical and weather data; satellite video teleconferencing networks to provide training and distance learning to the National Guard and reserve units and for telemedicine applications; and access-management approaches such as TDMA.

The ongoing synergy between the commercial and defense sectors is readily apparent in satellite communications. The introduction of commercial satellite communications in 1965 was limited to very small satellite payloads and required very large Earth stations to receive the very weak signals (e.g., INTELSAT Earth stations required antennas 100 feet in diameter). A 1971 experiment clearly demonstrated the feasibility of providing satellite communications—including digital voice, data, and fax services—to ships at sea. However, the business aspects of such services were not strong enough to justify the required investment in satellite and ground control systems. Subsequent events led to a contract between COMSAT, an international industry consortium, and the Navy to provide a GEO satellite system with an added commercial L-band package for ship-to-shore use. This agreement eventually led to INMARSAT, now widely used not only by large ships (e.g., tankers, cruise ships) but also by pleasure craft and mobile users around the world, who can transmit and receive data and voice via low-cost, briefcase-sized terminals. Terminals are also used on transoceanic airline routes for navigation, control, and passenger telephone calls. An INMARSAT spin-off, ICO, is building a MEO mobile telecommunications system using 12 satellites.

Similarly, the introduction of commercial DBS sparked military interest in developing the GBS to satisfy broadband data requirements in all environments, including the battlefield, ships, and logistics. The architectures of Ka-band (superhigh frequency, or SHF), high-speed interactive systems planned for commercial operation by the year 2000 will have an impact on the ultimate GBS design in the near future. Ultrasmall-aperture terminals in these systems will be able to transmit several megabits per second and receive 100 Mbps from a 24-satellite constellation. The GBS has been designed to leverage the current DBS satellites through modifications such as moveable spot beams and different frequency bands.

The opportunities for defense use of commercial off-the-shelf (COTS) products depend in part on the particular characteristics of a military operation. For the purpose of analyzing communications requirements, military activities can be divided into the following four categories:

- *Non-mission-critical operations* require general communications infrastructure to provide logistics, training, entertainment, and general administration. For these activities, the military has either purchased COTS systems or leased services operated over commercial carrier systems. For example, the Army Training and Doctrine Command uses a commercial teleconferencing system based on a VSAT to provide training to the National Guard as well as telemedicine services.
- *Limited peacekeeping missions,* such as the Bosnia deployment, can feature a mix of COTS communications equipment (primarily VSATs and video teleconferencing) and military systems.
- *Regional conflicts,* such as Desert Storm, can feature a mix of COTS and military systems depending on the threat to commercial assets. Conflicts of this type seem likely to benefit from the use of emerging COTS systems (or derivations) such as satellite-based personal communications and broadcast data satellites.
- *Strategic/global conflict* requires the use of "survivable" military communications systems whatever the cost, implying reduced use of COTS systems.

3.2 MOTIVATIONS FOR COMMERCIAL-DEFENSE SYNERGY

Two key factors currently motivate the DOD to seek commercial products and services. First, the size of the business and consumer markets and the nature of many commercial practices help achieve economies of scale at many levels. Second, commercial approaches to R&D reduce cycle time such that advances in technical performance can be integrated into field operations in a timely manner. The DOD therefore has both economic and functional reasons to adopt commercial products and approaches when they meet—or could be adapted to meet—defense communications requirements. The commercial equipment is likely to cost much less overall than would equivalent defense-unique systems. Furthermore, because commercial industry evolves very rapidly in response to a competitive marketplace, the DOD can leverage commercial developments to field equipment that offers advantages in size, weight, power, bandwidth, or performance much more rapidly than is possible using traditional defense procurement practices.

Economies of scale are extremely important in the development and deployment of commercial products. Firms seek a balance of cost and quantity when deciding whether and how to enter business or consumer markets. As examples, business products such as VSATs are built in

BOX 3-2
DirecTV Receivers: An Example of the Volume-Cost Relationship

A DirecTV receiver consists of an 18-inch antenna and a sophisticated mechanism for receiving a 40-Mbps, digitally mutiplexed data stream. With more than 1 million units sold in the first year, these receivers are among the fastest-growing new product lines in the consumer electronics industry. DirecTV receivers were introduced at a list price of $700; after 2.5 million units were sold, the price dropped by nearly 50 percent because of competitive market pressures and economies of scale.

Impressed by the capability of such a small receiver system, the Navy and other services determined that DirecTV technology could be adapted to meet the military's broadband transmission requirements. However, the quantity needed by the military—hundreds of terminals—is significantly smaller than the commercial market. If a DirecTV-like receiver were developed as a stand-alone military product, then the cost per unit might be hundreds of times higher than the commercial price because the development, tooling, and manufacturing-setup expenses would be amortized over a smaller production base and optimized for smaller production volumes. Three years after DirecTV was announced, the services were still working to define a military version. Had the features necessary to support military needs been considered before the product design was finalized, the DOD could have taken advantage of the cost reductions enabled by the market growth.

volumes of thousands per month, and consumer products such as DirecTV (see Box 3-2) are manufactured in quantities of hundreds of thousands per month. The cost-quantity relationship forced the semiconductor industry, which originally evolved to support military and space applications, to switch to a commercial focus. As shown in Figure 3-1, the commercial market for semiconductors soared, whereas the defense share declined. In 1975 worldwide military purchases of semiconductors totaled $700 million, approximately 17 percent of the global market (INSTAT/SIA Information Services, 1997). At that time all major semiconductor manufacturers had military-quality product lines, particularly for high-reliability and extreme-temperature applications. By 1995 the military share of the market had dropped to less than 1 percent (INSTAT/SIA Information Services, 1997). Most major semiconductor manufacturers have announced the phasing out or termination of military product lines. Now military contractors must use either commercially available parts or obsolete, but military-quality, semiconductor parts.

The commercial sector far outpaces the defense sector in production rates and volumes, not only for final products but also for subsystems and components. The largest DOD acquisition of communications equipment is the SINCGARS radio: The DOD has purchased 75,000 units over 10

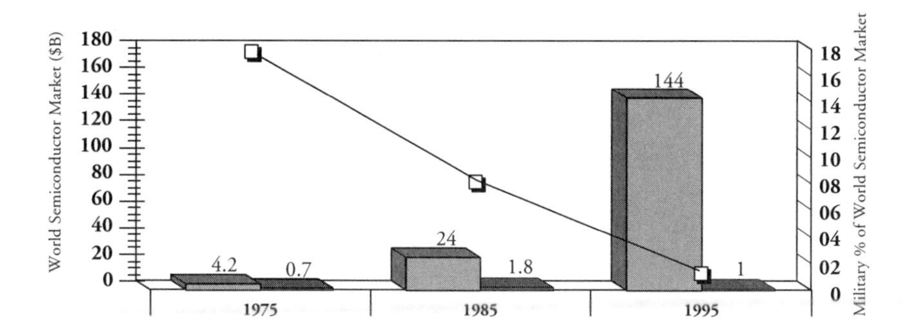

FIGURE 3-1 The world military share of the world semiconductor market dropped from 17 percent to 1 percent between 1975 and 1995 (the line graph, which corresponds to the scale on the right). During the same time period, the commercial market soared from $4.2 billion to $144 billion (the left bar graph for each decade) while the military market grew only slightly (the right bar graph for each decade). SOURCE: Joseph Neal, Commercial Plus Technology Operations, Motorola, Inc. Reproduced from Bradley (1996), with permission from the Semiconductor Industry Association and World Semiconductor Trade Statistics.

years of production. In contrast, commercial production of land-based mobile radios is in the range of 400,000 units per month, and cellular radios are produced in volumes exceeding 2.5 million units per month for the largest suppliers. To meet such market demands a typical cellular telephone factory might produce 5,000 telephones a day. Another factor distinguishing the two sectors is the open, competitive environment of commercial production. The market pressure for improved quality, pricing, and other features is felt by all commercial competitors, whereas the defense market has typically been limited to a few and sometimes just one contractor.

The next four subsections examine economies of scale manifested in several areas of commercial technology development: design, production, maintenance, and training. The fifth subsection reviews how cycle time can be reduced, thereby moving technical advances into the field quickly and also lowering costs over the life of a product.

3.2.1 Design Reuse

Commercial communications equipment typically is produced with a basic design that has a 2- to 5-year life span. The components used in that design are selected in the 1 or 2 years just prior to product introduction and typically represent the then-current state of the art in performance and cost effectiveness. Thus, during the product life the components remain cost-effective for the suppliers and are manufactured using state-of-the-art, cost-effective manufacturing facilities. Manufacturers typically anticipate new features in the market by modifying the design to use new components after 2 years of production. They also use components and manufacturing processes that are within a generation of the then-current state of the art, thereby operating close to the optimum level of cost effectiveness. By contrast, military equipment is often outdated: SINCGARS was designed more than 10 years ago, for example.

Commercial firms achieve the initial economies of scale through design reuse. The use of previous hardware and software designs can often save 50 to 80 percent of development time because detailed design documentation can be readily reproduced and design weaknesses can be largely eliminated using experience as a guide. Although the design cycle is not a major contributor to the economic cost of a commercial product, it is typically a large part of defense deployment cost. The DOD typically does not reuse hardware designs, instead relying on independent "stovepipe" systems, which are optimized to solve a specific problem. Some efforts have been made to create software libraries for reuse. Increased reliance on common building blocks could significantly reduce design cycle time (see Section 3.2.5).

3.2.2 Production Learning Curve

The learning curve is a statistical tool used to predict production costs and plan and control production. The curve is based on the assumption that there is a relationship between the time required to build a unit and the number of units that have been built; specifically, the learning process reduces the time needed to produce a unit as the cumulative number of units produced rises. It follows, then, that the less time it takes to build a unit, the lower the cost of that unit. If the cost of producing a unit follows an 80 percent learning curve, then there will be a 20 percent reduction in cost per unit each time the total number of units produced doubles. In the example shown in Figure 3-2, the first unit took 100 hours to build and the second unit took 80 hours, or 80 percent of the time and cost involved in building the first unit. The 10th unit required 48 hours, and the 20th unit required 80 percent of that effort, or 38 hours. The major factors that affect the cost of production are the initial cost, or the starting point of the curve, and the rate of improvement or learning, or the slope of the curve (Anderlohr, 1969).

The implication of the learning curve is that large volumes of standardized items, produced continuously (i.e., without significant hiatuses), reduce the cost per unit. Heeding this message, commercial production is fairly continuous, fluctuating somewhat with the demands of the market

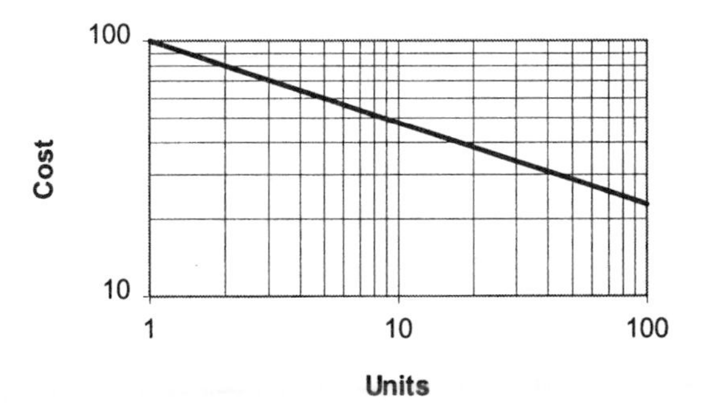

FIGURE 3-2 As more units are produced, manufacturing costs per unit decline steadily.

but generally changing processes gradually and with few interruptions. Although customized versions of products are increasingly in demand, the basic platform is usually consistent and the adaptations are minimal. By contrast, defense programs frequently begin with the building of only a few units, perhaps a few hundred, to determine feasibility or fulfill a limited need. Often these units are produced in numerous small batches with interruptions between the production cycles.

Cost management is practiced throughout the design and production of commercial products. For example, production volume typically needs to be known before detailed designs can be completed. The component costs, labor costs, and investments in labor-saving manufacturing devices are factored into the final design of a consumer electronic device. Large production volumes enable the manufacturing of designs that would not be viable at smaller volumes. A significant example is the fabrication of customized ICs with many functions that normally would be implemented in separate ICs. The large volume reduces the overall cost of components, parts, and assembly, even after the nonrecurring investments are taken into account. Another example is the design of electronic equipment for cost-effective manufacturing. These designs typically feature modules that snap together and minimal numbers of wiring bundles, fasteners, moving parts, and different part types. Costs are reduced further through incremental production changes. During the repetitive commercial design and production cycles, the boundaries between system, subsystem, unit, and component begin to blur as automation enables larger and larger subsystems to be treated as components. In this way, what was once a high-technology system (e.g., computer memory) becomes a commodity part.

Following the lead of the commercial sector, the DOD might achieve some economies of scale in production by revising its procurement practices to make large-volume purchases of basic COTS communications equipment for entire departments at one time. Some isolated efforts have been made in this regard, but there are ample opportunities to expand this approach.

3.2.3 Maintenance and Logistics Support

Economies of scale can be achieved in the maintenance of equipment after it has been developed and fielded. Equipment occasionally fails in the field because of design defects, manufacturing defects, worn-out mechanisms, lightning or power surges, or simply heavy use. Sometimes fielded equipment is upgraded during maintenance procedures to add new features or functions.

In the commercial sector field-failure data are typically analyzed on highly automated equipment, which can trace failures to specific modules and components and automatically update design and component history databases, including any links to environmental factors. Design updates are inserted into the manufacturing process throughout the commercial life of a product, thereby improving its robustness. After approximately one year of production, experience with field failures often has produced the feedback necessary to eliminate most design defects, reduce manufacturing defects to a level consistent with the current state of the art, and generally achieve the best product possible within price constraints.

Consumers rarely, if ever, pay for the maintenance or repair of low-cost communications equipment. Rather, warranties and service contracts are often viewed as a necessity in maintaining complex products that are not easily repaired; products are often replaced if they need repairs after the warranty expires. Viewing warranties as insurance policies, or guaranteed streams of income, specialty maintenance companies have emerged to provide a variety of maintenance tasks, both on site and at the factory.

In the defense sector, communications equipment is often maintained by the acquiring agency rather than the manufacturer, typically at greater expense. Typically a module is replaced and the equipment is retested, a strategy that usually finds the primary defect but sometimes misses marginal problems elsewhere. Large numbers of spare components need to be kept available, either in replacement modules or in component form such that modules can be manufactured, throughout the useful service life of the system. The supply of spares is often threatened when, because of the small production volume, the supplier no longer finds the component profitable to produce. When this occurs the manufacturer usually notifies customers of an "end-of-life buyout." The customers then try to project future needs and purchase enough components to satisfy them (not always at competitive prices). The DOD is known for keeping communications equipment in service far beyond the life span of equivalent commercial technology; typically, military systems are removed from service only after catastrophic failure. Defense equipment is often kept in use for 20 years, whereas component suppliers often set product lifetimes at less than 8 years; thus, the military needs to stockpile approximately 10 years' worth of components. Under normal circumstances, the DOD assumes that 25 percent of its equipment will need to be refurbished at some point.

The additional maintenance costs associated with traditional defense acquisition could be reduced if manufacturers—who can efficiently analyze all field failures, suggest redesign enhancements, and redesign components and modules to enhance cost effectiveness and other features—provided for maintenance and logistics support when appropriate. In addition, a reevaluation of military equipment maintenance practices may be warranted in light of the capabilities of advanced communications and transportation systems. Defense acquisition systems have historically provided logistics support for rapid equipment repair (i.e., within a few minutes of field failure) anywhere in the world by staging replacement modules near locations where critical equipment is in use. Such an approach may no longer be necessary.

3.2.4 Training

The commercial sector achieves additional economies of scale in training. The expense of user, logistics, and support training is built into the cost of new product introductions, and training is subsequently converted from expensive formats (i.e., personal, face-to-face support) to videotape, interactive CD-ROM manuals, on-line help, and literature. By contrast, the training of defense maintenance and logistics support personnel offers few economies of scale. Training materials are developed, but they are neither as detailed nor as widely distributed as are commercial manuals. Indeed, defense support training can remain somewhat diffused and superficial because the military uses so many types of equipment and relatively small numbers of each type. As the DOD purchases more COTS products, the use of commercial training materials might be appropriate.

3.2.5 Cycle Time

Commercial product design cycles, which usually last from one to four years, are set by competitive pressures: The first company to market a product with a new feature can reap large increases in market share and profitability.[2] A new commercial communications product is released every few months. Manufacturers therefore begin to field test and optimize the features of products before the designs are completed, accelerating the development process by several years. This environment fosters the introduction of new and improved technologies at a very rapid pace, often at a low incremental cost to consumers. Companies gain additional reductions in cycle time by designing products to accommodate new features on each new production run, often every six months.[3] These advanced commercial technologies are then available for rapid insertion into commercial or defense applications.

The military product design cycle is much slower. It begins when a contract is awarded and ends with delivery of the final product, which is not field tested or optimized until the design is completed. The developer does not have sufficient control over systems integration, testing, and evaluation to perform concurrent engineering that would reduce overall cycle time. Further delays are imposed because training, logistics, marketing, and distribution processes are not generally developed concurrently with manufacturing tooling equipment as they are in commercial systems.

A key design feature affecting cycle time is the ease of upgrading equipment. Commercial baseline products are designed to accommodate hardware and software extensions throughout the planned lifetime of the product. This is critical because of the high cost of the wireless infrastructure. The longevity of the infrastructure depends on a complex trade-off between the equipment offered by vendors and the pace of change in services. Typically service providers have a detailed road map that identifies when new services will be offered; these services are selected based on the equipment available at a reasonable cost and the market demand for a profitable service.

Commercial upgrades to accommodate new services and changes in market direction are generally implemented through software updates rather than more-costly hardware changes. Therefore, software plays a growing role in product development and cycle-time planning. Software is also often used to correct hardware problems, such as designs that were oversimplified to meet a price point. In such cases new software requirements are discovered late in the product development cycle, meaning that software is the last element to be developed and may be installed either just before or even after production. Yet software updates need to be thoroughly tested and tolerant of all environmental and loading factors. As a result, the software development process is now of great interest. The quality and timely release of software as well as software-defined infrastructure services are therefore becoming critical factors in the commercial communications industry.

Military radios, by contrast, are typically optimized to meet a single specification rather than designed for easy upgrading. Moreover, when upgrades are possible they usually can be implemented only by the original vendor. An exception is SpeakEASY, which was designed to support hardware and software extensions through its open architecture. Emerging military requirements (e.g., interoperability, open architecture) are driving the trend toward more flexible systems such as software radios. With the role of software expanding in both sectors, the commercial experience in software management may offer lessons for the DOD.

Both sectors are concerned about retaining the use of legacy equipment. As the complexity and cost of each new generation of infrastructure rise, the effects of obsolescence become more important to system planners, investors, and consumers. Concepts such as backward compatibility, while ill defined, have a very pragmatic meaning in the consumer electronics business: A technological advance should not result in denial of service to owners of older systems. Backward compatibility is also important to the military, which is likely to continue using legacy equipment for some time to come.

3.3 BARRIERS TO COMMERCIAL-DEFENSE SYNERGY

Although many defense communication needs can be met with commercially available equipment, a variety of barriers prevent a complete match across all systems. For example, commercial-defense synergy is not appropriate in the development of some highly specialized or classified command-and-control systems. There are organizational limitations as well. Traditional government acquisition has led to a large number of stovepipe systems, which are developed by a single contractor, meaning that other contractors cannot compete for follow-on development or production.

There are additional concerns regarding industry's current capability to meet the unique requirements of some military systems. In the wake of cutbacks in some procurement programs, many defense communications suppliers have begun to develop a commercial orientation to preserve their technology and manufacturing base. Commercial applications are attractive not only because of the vast market but also because they are not subject to federal and defense acquisition regulations: The supplier benefits from simplified accounting and has greater freedom to structure the details of contractual agreements. In the past, an increase in contracts was sufficient to rebuild capacity following a period of reduced defense spending. Now, however, action might be necessary to ensure that critical design and manufacturing capabilities are not lost altogether.

In view of the need to maintain surge capacity for those times when sudden military action is needed,[4] the declining industrial interest in military communications systems makes it all the more important to understand—and either overcome or accommodate—the barriers to commercial-defense synergy. These barriers include the risks of military dependence on the commercial sector, the contrasting approaches to making trade-offs between cost and complexity, and differences in communications infrastructures.

3.3.1 Risks of Dependence on Commercial Technologies

The DOD assumes certain risks when using commercial technologies. Most importantly, significant military use of commercial technology increases the possibility that a potential adversary could have or gain detailed knowledge of the systems. In the case of computer networks, the defense infrastructure is built almost entirely of widely available computers, software, and networking components. An adversary could exploit the known weaknesses of these components, a possibility that has prompted efforts to improve network robustness. Any weaknesses in the commercial communications infrastructure might be vulnerable to similar exploitation.

One solution is to use communications equipment that is unavailable to potential adversaries. Such products include those subject to U.S. export controls, which are intended to keep certain advanced technology products and weapons systems away from designated hostile countries. Several types of advanced commercial communications technologies, including those involving extensive computations or spread-spectrum techniques, fall into this category. It is difficult to gauge the impact of international technical awareness of military effectiveness because technology is only one of several influences (the others include strategy, tactics, training, and weaponry). Critics of export controls contend that many affected technologies are readily available from offshore sources and that the controls serve only to reduce sales for U.S. manufacturers and provide evidence of who is acquiring these commodities.[5] But with the rapid evolution of commercial wireless technologies, especially software-based systems that could be converted to implement new waveforms or military countermeasure capabilities, there is also an argument for strengthening controls on the export of certain advanced communications technologies. Although the issue is beyond the purview of this committee, a review of export controls may be warranted.

Another risk of commercial dependence is the sometimes-hidden difficulty of making what seem initially to be simple modifications to COTS systems. End-to-end encryption or AJ technologies, for example, might be added to a commercial system to meet military security needs. However, given the high commercial production rates, such modifications are often very difficult to implement. For example, the key interface signals or connections could be embedded in an IC and unavailable for specialized wired connection to an add-on feature. The safest strategy for the DOD is to use a common baseline technology and common components but pursue a separate design effort. In addition, the DOD could participate in standards-setting activities to encourage the development of baseline commercial equipment designed to accommodate the addition of militarily useful features.

A third concern is the availability of commercial systems in wartime. Many of the largest satellite networks, including those shared by the U.S. military, are owned or operated by international consortia. Although these providers are reliable partners in peacetime, whether they would give DOD priority or expanded bandwidth during times of conflict is unclear. Such preferential service could be hampered by the operators' need to serve other customers or possible unwillingness to provide support in a controversial conflict.

3.3.2 Trade-offs Between Cost and Complexity

Perhaps the most obvious barrier to direct commercial-defense synergy lies in the contrasting strategies used to make trade-offs between system cost and complexity. As a potential user of commercial services the DOD has certain expectations, many of which are requirements if it is to fulfill its mission. In a number of dimensions, these expectations are at odds with the criteria used by commercial communications services in designing and deploying their products. Simply stated, the military has some extraordinary needs, whereas the commercial sector tends to focus on delivering reliable but ordinary service.

3.3.2.1 Performance Issues

For example, the military expects to use leading-edge technology. The present analysis is based on the assumption that the DOD cannot achieve its mission with technology that is inferior to that of an adversary. It is also assumed that every adversary's technology is state of the art. But commercial communications services are rarely based on the most advanced technology available. Rather, providers deploy technology based primarily on its cost effectiveness and affordability, that is, whether customers are willing to pay for the capability.[6] Over time, production volumes increase and costs decline, but the initial costs of an advanced technology can be a barrier to its commercial application.

The DOD also requires that certain functional capabilities (assuming they are technologically feasible) be deployed in any location where the military needs to operate. Military operations are extraordinary events in which communications traffic is unpredictable, driven by the characteristics of the individual operation. Yet the cost of an inability to communicate can be very high, and so the probability of such a breakdown needs to be kept very low. By contrast, commercial communications providers have limited resources and cannot cover every service area that could be profitable. Commercial systems are designed to serve a particular area for many years with slow changes in technologies, features, and volumes. To minimize life-cycle costs, such systems are based on fixed facilities that cannot be deployed or shifted rapidly to meet an immediate demand. Commercial wireless systems are generally engineered to meet the peak traffic requirements of the average business day. They are not designed to meet the requirements of extraordinary events under emergency conditions, even regular and predictable ones.[7]

Similarly, commercial and defense equipment differ in their tolerance for unusual environmental conditions. Military units are likely to encounter extreme environments such as jungles, deserts, or polar regions and be subjected to the harsh conditions of battle. Therefore, regardless of the added cost and system complexity, defense communications equipment needs to be designed and built to tolerate extreme temperatures, submersion, high levels of shock, explosions, and vibration. By contrast, commercial manufacturers and consumers are unlikely to incur additional expenses for equipment that operates under extreme conditions. In fact, consumers often favor the least expensive product over one with the best performance, warranty, survivability, and advanced features. Consumer products such as telephones are designed to survive reasonable levels of wear and tear and perform under moderate environmental conditions.

3.3.2.2 *Quality and Testing*

Military communications equipment can be highly complex and pose difficult testing and diagnostic challenges. For example, a military system can encompass networks of computers, each running a real-time suite of applications in support of a system-level application. When a safety or mission-critical function is involved, elaborate procedures are followed to provide for multiple redundancies, failure detection, independent software development, and cross-checking to ensure reliability. But exhaustive testing to identify problems that have an extremely low probability of occurrence (e.g., once every few hundred thousand instructions) may not be cost-effective if there are no safety implications or critical effects on performance. In cases that may not justify exhaustive testing, new work in formal methods (i.e., mathematical techniques that obviate the need to test every possible situation) holds promise for further reducing the likelihood of rare anomalies.

There is a widespread perception that defense equipment is tested much more thoroughly than are commercial products. In truth, commercial testing procedures vary widely but can be quite rigorous. For example, manufacturers of industrial-grade communications products have largely adopted high levels of integrated quality control. In the semiconductor industry, firms expend significant resources even after a chip is designed to create an exhaustive test pattern capable of catching both design and manufacturing flaws.[8] Many vendors have adopted processes that are comparable in function to those required for defense equipment, although the conditions are likely to be less extreme.[9] Likewise, commercial electronics systems have improved in recent years because consumers now demand unprecedented quality and reliability, even under environmental conditions that just a few years ago would have been considered strictly military grade. Table 3-1 displays the stringent test parameters for one modern commercial product, a car radio. Drivers now view the performance of a radio as reflecting on the quality of the vehicle. As a result, most of the environmental qualification tests for a car radio are actually comparable to tests conducted on a high-performance Navy jet fighter.

As a result of these trends, some commercial products may be tested thoroughly enough to meet defense needs without further testing under military conditions. In some cases, it may not be cost-effective for the military to test and inspect commercial components already shown to have very low failure rates. This is most likely to be the case for equipment that has few moving parts and is designed for human use under extreme conditions.

TABLE 3-1 Commercial and Military Testing Parameters

Environmental Test Parameter	Delco AM/FM Radio	Hughes F-14 Radar
Low-temperature storage	–40 °C	–62 °C
High-temperature storage	85 °C	95 °C
High-temperature endurance	70 °C	Not tested
Still-air operation	35 °C	Not tested
Low-temperature operation	–40 °C	–18 °C
High-temperature operation	70 °C	71 °C
Random vibration	0.05 *G* per hertz	0.015 *G* per hertz
Shock	20 *G* per 11 milliseconds	15 *G* per 11 milliseconds
Humidity	90% @ 65 °C	85% @ 71 °C
Power/temperature cycling	1,000 cycles	Not tested
Explosion	Not tested	Tested

3.3.3 Infrastructure Differences

The differing expectations of defense and commercial customers affect communications infrastructures. Cellular customers want to be able to make calls whenever they wish. Military users have much more complex needs for security, LPD/I capabilities, concealment of network functions, message priority and preemption, tolerance of severe environmental conditions, and, perhaps most crucially, the capability to work without any fixed infrastructure. These needs generally preclude the use of antennas on fixed towers, highly engineered site installations, site-specific antenna selection, microwave links to central switching offices, and preplanned routing tables for switches. (Exceptions include permanent posts and fortifications and a considerable amount of mobile infrastructure mounted on trucks, trains, and large aircraft.)

In operations other than war, the use of existing local communications infrastructure (where available) can offer tremendous cost and performance advantages. It remains crucial, however, to provide military-grade security when using cellular communications or other wireless data services; such security measures need to be interoperable with the U.S.-based military communications infrastructure, including the Secure Terminal Unit III (STU III; the government standard in secure telephony) and related systems, and the array of secure networking products (e.g., the network encryption system [NES] equipment for strategic defense networks).[10] The DOD is addressing these issues through the CONDOR program, which is developing a cryptography module for cellular telephones used by the military (commercial versions are already on the market, and additional spin-offs are likely). The system will be interoperable with STU III. The security of local infrastructures remains an issue, however, as does the continuing need for interoperability among defense networks.[11]

In regions lacking a local communications infrastructure, substitutes will soon be available. The Iridium, Teledesic, and GBS satellite systems will provide worldwide communications. The military could also use unmanned aerial vehicles (UAVs; special aircraft that can fly along a programmed set of waypoints and perform programmed tasks) as high-altitude platforms for communications and reconnaissance services. Among their advantages, UAVs weigh less than manned aircraft (leaving more room for payload) and have no need to carry life-support systems such as oxygen and emergency ejection equipment. The UAVs could serve as relays between wideband satellite links and battlefield communications systems, carry sensors, jam or otherwise attack enemy information systems, and even provide positioning information in the event that the GPS is shut down during hostilities to prevent its use by adversaries.

The high-altitude endurance system (a UAV-based airborne sensor system) and the airborne communications node (ACN)[12] UAV-based platform are designed to be cost-effective, multipurpose platforms delivering C^4I capabilities.

3.4 DESIGNING WIRELESS SYSTEMS FOR MILITARY APPLICATIONS

As noted previously in this report, some features required in defense communications equipment are not generally available in standard communications products and could be difficult to add. Some advanced features, such as LPD/I and AJ waveforms, interoperability with the DOD's nearly two dozen legacy waveforms, or highly specialized spreading waveforms, may be attainable only in specialized military systems such as the SpeakEASY software radio. The following sections discuss the design of future wireless systems for defense applications, focusing on three key issues: network architecture; security; and multimode, multiband systems. Some commercial products could be adapted to meet defense needs in these areas, but specialized military research will likely be needed.

3.4.1 Network Architecture

3.4.1.1 Network Design Issues

Most commercial communications infrastructures use a base-station-oriented architecture, in which communication flows between a base station (equipped with a well-sited antenna and a high-power transmitter) and wireless terminals. Cellular, paging, trunked radio, and various data radio services use this model. This design is not entirely appropriate for military communications infrastructures, which need to allow for immediate deployment with no set-up time, no siting advantages, minimal antenna advantages, and continuous movement of all network participants.

The DARPA GloMo program has focused on peer-to-peer, multihop packet radio networks. However, there are valid questions about the suitability of a peer-to-peer model for military command-and-control hierarchies as well as the RF link penalty paid by this design.[13] The GloMo perspective was expanded recently to include other architectures, but the program has never assessed all the possible choices, even though network architecture forms the basis of any telecommunications system. Among the issues to be addressed are peer-to-peer versus base-station-oriented design, connection-oriented versus connection-free architecture, single-channel (e.g., Ethernet) versus two-channel structures, bandwidth issues, and protocol selection.

For example, the choice of a connection-free architecture is particularly important for applications that are intermittent in content, as is typically the case in battlefield situations. (The inefficiencies of a connection-oriented architecture are discussed in Chapter 2, Sections 2.1.6.2 and 2.1.6.3.) The choice of a two-channel architecture can be particularly important if, as is often the case in military applications, much more data are delivered to a client than are received from the client. This design enables maps and situational-awareness data, for example, to be disseminated in a battlefield without any multiple-access penalty to an unlimited number of receivers. Meanwhile, location information, such as identification and coordinates, can be sent from many field areas in short, spread-spectrum bursts over a separate multiple-access channel.

Numerous research issues need to be resolved before a military topology with optimum performance and overhead can be defined. Continuing R&D and demonstration projects over the next decade will help define the basis for commercially successful standards, which may or may not suit military needs. Current commercial systems treat each application separately, whereas the military probably needs to take an integrated approach. However, many commercial designs, such as the following eight examples, could have military applications. These designs would need to be analyzed and perhaps modified before their application to military systems. Even so, the effort would likely yield a considerable savings to the DOD.

Enterprise Networks. Enterprise networks are being deployed worldwide based on standards such as ATM, X.25 (a standard interface for packet network access), frame relay (a potential successor to X.25), and TCP/IP. The products include comprehensive WAN solutions that encompass multiple technologies such as LAN/WAN interconnections, dynamic routing, accounting, statistical information, and performance monitoring. Some new architectures unite connection-oriented WANs with connection-free LANs over both narrow and broadband channels. Critical questions remain to be answered about the suitability of this technology in the battlefield environment; these are issues that could be addressed as part of a broader assessment of military network architectures. System planners need to minimize the vulnerability of centralized control points and determine the bandwidth required to distribute routing information updates on degraded channels as units move around the battlefield.

Cellular Telephone Systems. Commercial cellular networks consist of both analog and digital systems conforming to various standards. With cell sizes of up to 10 km, these systems can cover a broad area, sometimes including wireless local loop service. The technical feasibility of using cellular systems in a battlefield environment requires further study. For example, cellular systems introduce special complexities because of the need to hand off transmissions between cells. Performing a handoff when system coverage may be incomplete over the theater is an elaborate process; equally complex are the rules for assessing when a handoff might be advantageous in light of possible jamming. Furthermore, mechanisms still need to be developed for authorizing access in a mobile tactical network. In cellular systems, these activities are performed in the cellular mobile switching office using the home location register.

Low-Tier Systems. In contrast to cellular systems, low-tier systems use low-power microcells of up to a few hundred meters in radius, small rather than tower-mounted antennas, and 32,000-bps voice coding for high quality and low delay. These systems are designed to serve users moving at pedestrian speeds. They are also suited to wireless local-loop applications because the round-trip delay is under 2 milliseconds, the quality of speech is comparable to that for wired services, and the short distances between hub stations and users generally result in low fade levels. Given these features, low-tier technologies could play a role in military communications.

Authentication and roaming capabilities are also provided in current-generation PACS and PHS systems. Packet-transfer protocols are being developed that will enable low-tier systems to serve as the transport mechanism for a wireless LAN while simultaneously carrying voice traffic. Commercially, the most successful low-tier system is PHS, which serves several million subscribers in Japan. The PACS system, developed in the United States, has not attracted significant markets to date but is the focus of active R&D efforts. Areas of investigation include techniques to incorporate flexible antennas and channel equalization to extend the range of PACS systems, especially to high-speed mobile applications. To determine the applicability of PACS and other low-tier systems in the battlefield environment, system planners need to address various issues, including security, antennas, adaptive waveforms, and operating range.

Radio LANs. The PACS system provides a protocol that is ideal for wireless LAN applications. The PACS packet channel (PPC) protocol provides the user with a variable bandwidth and asynchronous, asymmetrical data service at rates up to 256 kbps per radio port. The PPC converts the physical layer of PACS from a circuit-switched protocol to a packet-switched protocol consistent with TCP/IP. The current voice-PACS architecture is capable of circuit-switched connections, with a radio port control unit at the hub providing connectivity into the public telephone system. The radio LAN PACS has a distributed hub architecture, appearing to the user like a device connected to a wired LAN, when in fact it is connected to a radio transmitter. A radio port can be configured to provide packet data on some time slots and voice on other time slots. Radio ports are small units that are easy to install, require no special towers, and can support up to 238 users each. The LAN PACS system can provide a quick, efficient means of installing a wireless LAN. The PACS technology is an open standard, which makes it easy to obtain equipment.

Local Multipoint Distribution System and Related Technologies. Ka-band (SHF) frequencies are now being used in wireless cable connections linking end users, urban fiber loops, and local telecommunications bypass companies. The frequencies involved are the 18–19 GHz band, known as the digital electronic message exchange; the 27.5–30 GHz band, known as the local multipoint distribution system (LMDS), which is used in some locations as a one-way television delivery system ("wireless cable"); and the 38–40 GHz band, which can be used for line-of-sight (LOS) data transmission at rates of up to 155 Mbps. The last band is used extensively in Europe for backhaul transmissions of personal-communications signals from base stations to mobile switching centers. Because LMDS and related systems offer high throughput, they could be useful to the military in short-range (approximately 5 km), nonmobile elements of untethered communications systems. Further study is required, however, because the system would require a high SNR to be demodulated correctly and might not be suitable in the presence of jamming.

Line-of-Sight Relays. The introduction of wireless cable systems has led to the development of high-speed LOS relays. These devices provide wideband access through a remote hub in areas lacking direct LOS access to a primary hub station. The relays generally operate at low power because the antennas pointing in both directions (i.e., toward the primary hub and the users) are parabolic dishes with very narrow beam widths rather than the sectorized antennas used in wireless cable systems. Among other advantages, LOS relays operate continuously and therefore do not require burst-mode modulation. Systems based on relays are actually simpler in design than are wireless point-to-multipoint systems.

Satellite Networks. The DOD currently takes advantage of VSAT and telephony Earth stations for information gathering and two-way, transaction-oriented traffic. Approximately 40 hubs exist around the world, which could be used to backhaul PACS or cellular wireless nets deployed in the field. Regional and global mobile-personal-communications systems (e.g., Iridium, ICO, Globalstar) are being developed for deployment around the year 2000. These systems are being designed to operate in either L-band (UHF) or S-band (UHF/SHF) frequencies. Broadband Ka-band satellites (e.g., Teledesic) are being designed to use high-speed, interactive, low-cost Earth terminals.

Satellite traffic service levels for LEO and MEO systems tend to run an average of 0.05 to 0.2 erlangs (i.e., the channel is occupied 20 percent of the time) per square kilometer. Planned systems are expected to have spot-beam capabilities that will increase traffic service levels by several orders of magnitude, but DOD battlefield communications needs will likely exceed the service level of any single system. Because the link margin varies with the square of the range, shorter links provide greater advantage. Therefore, battlefield communications services will need to be supported with a hierarchy of links that graduate in altitude according to the transmission distance. Satellite communications will play a role but many types of systems will be required, including handheld units, vehicular and aeronautical platforms, and high-altitude UAVs.

Network Management Systems. Network managers enable a central controller to monitor and change system parameters using standard software, graphical user interfaces, and relational databases. The controller isolates faults, produces status summaries, keeps usage statistics, and changes configurations. As these functions become standardized, the DOD could adopt commercial network management systems as a means of simplifying enterprise network operations.

3.4.1.2 Bandwidth Requirements

Battlefield communications currently consist mostly of voice and a very limited amount of text-message traffic. Communications equipment is not broadly available to individual soldiers below the noncommissioned officer ranks.[14] Approximately 10 percent of soldiers now have voice communications, and only satellites, certain aircraft, and smart missiles carry sensors for still imagery or video.

The digitized battlefield of the future is based on the concept, verified by the Gulf War, that extensive real-time data gathering and surveillance can improve situational awareness and battle management. Military commanders want an accurate, real-time image of the total battlefield that indicates the positions of friendly and enemy forces; provides still and video images as well as data from infrared, radar, and other sensors; and is integrated with communications systems to ensure that the right information is distributed wherever needed. The realization of this vision will require significantly higher bandwidth, both on a link-by-link basis and in the aggregate (bits per second per cubic kilometer), than is now possible on the battlefield. Indeed, assuming that all future ground troops are equipped with communications and image sensors, there will be order-of-magnitude increases in both the numbers of users and the bandwidth required for imagery, as the data will flow higher and farther in the military hierarchy than voice typically does. Moreover, users will not be willing to spend half an hour delivering a single image, nor will their batteries tolerate such loads. New systems will be needed to provide higher bandwidth and hierarchically sensitive store-and-forward networking that will minimize the power required to transmit wideband signals.

The commercial sector has focused primarily on providing as many narrowband voice channels as possible. Wideband applications such as remote surveillance systems are becoming common in the industrial sector, and video-on-demand and video games are potential commercial markets. But unless substantial markets emerge for high-bandwidth services, the commercial sector will be slow to produce high-bandwidth wireless communications products that exploit the results of the third-generation R&D efforts described in Chapter 1. Military system planners have one advantage over their commercial counterparts in that mobile soldiers often operate within a short distance of a vehicle. In the digitized battlefield concept, many military vehicles could be equipped with radio equipment that can serve as high-power repeaters and may have very-high-gain antennas relative to those on the handheld units. Although practical levels of transmit power and antenna gain depend on the carrier frequency used, the availability of well-equipped vehicle platforms will make it feasible to transmit imagery to and from soldiers with handheld units.

The DOD's combined requirements for real-time data traffic and high bandwidth suggest that ATM technology, which is expected to be popular commercially, might be appropriate for battlefield systems, at least within the wired network. This technology operates at high bit rates through fiber-optic interconnections. Fixed-size data units (53-byte cells) facilitate efficient switching while connection-based semantics (i.e., virtual circuits) enable bandwidth to be allocated before a connection route is established. In addition, ATM switches can provide sophisticated mechanisms for choosing how to multiplex flows, thus offering a variety of QoS levels for different types of traffic. Finally, ATM supports message priority, queue management, and admission control mechanisms to yield performance guarantees.

The seamless integration of ATM and TCP/IP technology into the wireless battlefield communications architecture would not be a trivial undertaking.[15] Gateway functions will likely be provided at hierarchical RAPs, nodes that provide a variety of cross-networking, repeater, and other information services. The implementation issues are widely debated in the network community (and extensively studied in the ATM

Forum) because ATM's mechanisms for end-to-end QoS cannot be supported directly by current Internet protocols. An ATM-only solution is also problematic for many reasons, most significantly because it is impossible to guarantee QoS for wireless systems in motion (the link margin is continually changing and the signal can be lost completely).[16] Thus, wireless-link-level protocols of increased sophistication will be needed to ensure that data can be delivered successfully across the link while still meeting the QoS guarantees given during connection setup. The commercial sector is likely to resolve these technical problems eventually, but the DOD would do well to stay close to these debates to ensure that its interests are represented. Support for research and prototype development, coupled with testing of emerging technologies in battlefield exercises, could be useful government roles.

The use of GBS services will enable the direct delivery of wideband services to the battlefield and provide terminal integration opportunities for the "smart push, smart pull" concept (in which the warfighter receives customized data in a timely fashion without expending much effort to define the needs). Real-time, in-theater sensors will provide more recent information and more immediate task assignments than will DOD's satellite assets; thus the realistic information flow model is within the battlefield, with copies back to the Pentagon.[17] In the Army's vision for the digitized battlefield of the future, bandwidth is allocated not only up and down the command hierarchy but also horizontally to cooperating formations.

3.4.1.3 Source Coding

In current voice coding technology, speech is compressed with many different compression algorithms to bit rates ranging from 300 to 64,000 bps. Voice quality and compression factors have improved over two decades of research to the point that, in moving voice traffic, linear predictive coding (LPC) and other voice coding technologies enable compression ratios of 26 to 1. Even greater compression is possible to provide additional LPI and AJ advantages but at the cost of reduced speech quality and increased delay. Encoding of data has also been reduced to a well-known process, thereby providing a standardized method (e.g., using the Lempel-Ziv algorithm [Ziv and Lempel, 1978]) for both commercial and defense applications.

There is now great interest in coding of not only voice but also sound, images, and video. Reasonable-quality compressed video suitable for conferencing applications can be achieved with video encoders and decoders operating at 64 to 128 kbps, the equivalent of a dozen digital voice channels in commercial cellular systems. Videoconferencing will require substantially higher bandwidth and will need to demonstrate operational effectiveness if it is to be used extensively on the battlefield. Similarly, high-resolution still or sensor images (e.g., a small image of 480 × 640 pixels at 30 bits per pixel) will require representations of well over a megabyte, far exceeding the capabilities of existing tactical radios such as SINCGARS, which transmits data at about 2.4 kbps (requiring an hour to transmit the image just described).

Standards are under development and chip sets for standardized implementation have recently become available for defense and commercial purposes. The most notable standards are the joint picture experts group (JPEG) and motion picture experts group (MPEG), which encode images, including the high-energy, low-frequency components. Image coding was developed for military reconnaissance but is also used in full-scale video teleconferencing; many military users are mobile and communicate through wireless links, whereas commercial users are stationary and connected by fixed, high-quality transmission systems. (Commercial-defense synergy is a tradition in source coding, as described in Box 3-3.) Image coding and video source coding technologies can compress a typical still image by a ratio of up to 100 to 1; newer technologies offer up to four times the compression of deployed systems.

BOX 3-3
Commercial-Defense Synergy in Source Coding

Source coding technologies have often been shared by the commercial and defense sectors. An early example was Sigsaly, a World War II voice communications system that relied on a tractor trailer full of security equipment (which today would be referred to as a channel bank "vocoder"). After the war, AT&T explored many commercial spin-offs of this technology, leading to the spectrograph and several technologies for voice coding. Source coding became more sophisticated as semiconductor technology replaced vacuum tubes and greater functionality was achieved in a smaller package and at lower cost. One result was continuously variable slope delta modulation, widely adopted in defense and civil government applications for secure communications. The next major step was the development of linear predictive coding (LPC), which uses a mathematical model of a voice signal and enables speech to be represented at 2,400 bps, a rate low enough to be combined with available modems to provide real-time, secure voice communications over dial-up telephone lines. This discovery was widely used in secure defense communication in the form of the STU III telephone, which continues to feature this data rate. Extensions to the LPC source coding technology developed to the point that the speech quality at low bit rates was acceptable for commercial purposes. These approaches were adopted in several digital cellular communication standards, combined with multiple access techniques, and used to develop unique standards.

Compression ratios and image quality are likely to improve with time, but the evolution of the commercial technologies has been constrained by the wide proliferation of the JPEG and MPEG standards and the tools built around these standards. Furthermore, commercial data and video coding standards have not yet evolved to be robust in the presence of the bit errors introduced by wireless transmissions. Continued improvements in source coding technologies are needed so that the bit rate required for faithful reproduction of information can be limited and the DOD can control the growth of its bandwidth requirements. An additional concern is the vulnerability of commercial source codes to jamming, an issue that could be addressed by research on adaptive waveforms.

The extent to which the DOD will use compression and decompression techniques to reduce its bandwidth requirements is unknown, and so it is difficult to predict the impact of advanced data file compression. However, because imagery represents a large component of the data traffic supporting defense activities, at least some benefits are likely. There would be some costs involved: New software would be needed for workstations, and new compression technology would need to be deployed into sensor assets.[18] But as the use of advanced collaborative planning and intelligent databases grows, specialized or localized compression strategies on database file transfers will help control the pressure for additional communications bandwidth.

3.4.1.4 Highly Adaptive Systems

Advanced modulation and smart radio technologies offer the promise of flexible, dynamically changing communications systems that can adapt to almost any conditions. Such systems will be able to select modulation, spreading code, and FEC and interleaving algorithms that will perform optimally in any environment—even in the presence of noise, jamming, or interference—while also meeting QoS requirements for delay as well as data rate requirements. When applications can detect prevailing channel conditions and the radio system and external networking options can support adaptation, the user can gain at least an order-of-magnitude improvement in range, bandwidth, and AJ or LPI capabilities.

Ongoing research in this area has several shortcomings from the military perspective. The commercial sector is pursuing extensive R&D but its interest in adaptive systems is motivated by profit (e.g., accommodating more users per hertz), whereas the military seeks functional advantages such as increased interoperability and AJ capabilities. The DOD also supports research on adaptive systems, in part through the GloMo program, but the technologies are generally not demonstrated and tested under military conditions.

For example, existing network protocols are generally designed for static configurations and high-quality broadband links. In packet radio networks, discovery algorithms (which identify neighbors as each radio moves about on a network) determine the proper store-and-forward sequence for moving communications traffic toward specific destinations. These algorithms are now being developed, assessed, and standardized. To minimize overhead and streamline the discovery process, routing information servers are provided by the network that mobile units can query to determine who is connected and select the latest optimized routing paths to a specific destination. These algorithms have not been tested in military networks where mobility and network survivability under degraded channel conditions are of primary importance. The relative amount of overhead in a highly dynamic environment—including degraded channels—needs to be modeled in more detail.

Military systems will benefit further if source coding, cryptography, and antenna beam performance can also be adapted to prevailing channel conditions. The TCP/IP suite tries to deliver data without errors. In wireless military applications it is likely to be preferable to trade off bit errors (and the delay tolerated by the application) against the quality of representation of the original information. In voice applications, for example, users are often willing to tolerate occasional distortions of isolated words (but not entire sentences) but are intolerant of delay. Reconnaissance systems may require that no transmission errors will be acceptable, whereas other image delivery systems may allow a moderate number of localized image distortions, in exchange for more rapid delivery.

3.4.2 Security

Communications and network security have attracted attention recently under the umbrella of information warfare issues. Enhancements can be made readily in the wired infrastructure but are more difficult in wireless networks, especially packet radio systems for which protocols are still in the formative stages. Commercial and defense communications networks face different threats. Commercial providers are most concerned about the fraudulent use (theft) of service, whereas users of these systems are most anxious about improper access to and manipulation of their data. The military, which is often both provider and user, seeks to protect all aspects of its communications—not only the message but also the source and destination information, the inner workings of its equipment, and even the existence of a network.

3.4.2.1 *Availability of Service*

The commercial and military sectors have different concerns with respect to service availability. For commercial systems the primary issue is interference among users. This type of interference is well modeled, usually has stationary properties, and can be countered with thoroughly studied solutions. Hostile jamming in a military conflict creates a totally different type of interference, which cannot be mitigated using ordinary approaches. Hostile jamming can create a situation in which no usable, undistorted parts of a message are received. Other forms of jamming include intentional disruption of key information bits in messages, playback of old messages that are no longer relevant, or transmission of noise sequences to trigger false receiver actions.

A determined electronic attack on a military communications network could not be countered by any existing commercial equipment or any simple modifications to such equipment. Furthermore, it is unlikely that any future commercial system could satisfy military AJ requirements to the degree offered by defense-unique systems. Current military systems with AJ capability include the SINCGARS and Have Quick radios and MILSTAR satellite system. Because equipment used for jamming is becoming easier for potential adversaries to obtain, there is a growing need for development of advanced AJ techniques such as nulling and scanning antennas, spread-spectrum modulation, approved secure-spreading codes, elaborate error detection and correction, time-stamped messages, adaptive jammer-sensing techniques, and adaptive jammer-responding modems.

Access to commercial communications networks generally cannot be denied to segments of the population that have the proper equipment and, if necessary, are willing to pay for service. This feature is unattractive to military planners, who would prefer that communications systems offer normal service to friendly customers while blocking access by adversaries. Weather broadcasting, for example, is a vital part of tactical planning, yet most standard commercial systems cannot simultaneously guarantee the delivery of weather reports to critical commercial services while denying such reports to the enemy. A variety of technical approaches, similar to those used to control access to DirecTV and DirecPC, are available to safeguard broadcast digital data. Cryptographic codes and secret information exchanged in advance can enable selective access to some broadcast information.

3.4.2.2 *Confidentiality and Integrity*

Defense systems secure both the message data and, in a separate process, the routing information. Existing security systems cannot prevent an adversary from detecting the presence of military communications. Information about the battlefield, important targets, and plans of interest can be inferred based on the volume of communications to and from particular locations, and much can be learned about routine military activities from the modeling of communications traffic. As stealth aircraft, ships, and other vehicles are deployed, it is critical to guard against the detection of these platforms based on their radio communications.

The DOD uses LPD/I systems to hide the evidence of radio transmissions. These systems rely on unusual transmission frequencies, spread-spectrum techniques, narrow-beam antennas, low-power transmission, or very brief messages. By reducing an adversary's awareness of transmissions, LPD/I systems also minimize jamming efforts and their impact. Communications systems with these features can be of great value in many military applications because they deprive adversaries of information about deployment of troops, routine versus unusual operations, the communications hierarchy, and level of alertness or activity. However, advanced LPD/I communications techniques have not been strongly supported in recent years.

Commercial systems offer considerably less protection. For example, cellular communications can be easily detected, jammed, or demodulated and user location can be pinpointed readily. Equally vulnerable are commercial CDMA systems, which are based on direct-sequence spread spectrum with published spreading codes and do not provide LPD/I capabilities. Current digital wireless standards make provisions for privacy and authentication to block unauthorized use, but cellular wireless standards make no provisions for traffic security, meaning that information about the routing, content, and significance of the data can be intercepted. Signaling information is sent in the clear, although the identity of the caller is protected by an identification number known only to the user and the system (or is temporarily assigned by the system if the user is roaming to a new location). In packet-switching protocols, ATM, frame relay (an interface for packet network access), and fiber-optic networks, all addressing and signaling information is left unprotected so that switching or routing equipment can also read and interpret addressing, routing priorities, and other information contained in the headers.

Commercial service providers are not expected to expend significant resources to harden commercial infrastructures against attack, although they want to prevent losses resulting from fraud. Careful design strategies will be required to deter both fraudulent use and attacks, threats that will be nearly indistinguishable on a mobile packet data network. The growing use of such networks will motivate commercial research aimed at solving these problems. Yet even if the commercial sector achieves significant security advances, the DOD is likely to prefer at least some of its own approaches because COTS technologies would be readily available to adversaries.

One area where military R&D could be helpful is encryption, which protects voice, data, or video in an information frame or packet. A typical encryption system for real-time communications involves a cipher that encrypts the relevant bit stream one bit at a time. The decryption process requires both the relevant decryption key as well as synchronization between the sending and receiving process. (Synchronization refers to the assurance that a particular bit being decrypted in fact corresponds to the bit that was originally encrypted and sent to the receiver.) In principle, synchronization requires only knowledge of the starting point of the incoming bit stream, but in practice, establishing and maintaining synchronization throughout the duration of the transmission is complex. Current cryptographic systems are often inefficient because the synchronization consumes bandwidth and because synchronization may be performed packet by packet. (Any user of the STU-III secure telephone system can attest to the time it takes to achieve end-to-end synchronization.) The design of improved synchronization algorithms would streamline the security system and free up bandwidth for other uses.

3.4.3 Multimode, Multiband Communications

Interoperability has long been a goal of military systems. There are more than 17 different U.S. defense communications networks, and the sharing of messages among them requires the deployment of many unique information gateways or bridges. Furthermore, compatibility among U.S., North Atlantic Treaty Organization (NATO), European, and United Nations systems is increasingly important to military operations. For example, one report on the Bosnia Implementation Force notes that close air support missions can involve British Harriers, NATO E3-A (airborne warning and control system) aircraft, Norwegian forward air controllers, and Swedish-led brigades (Allard, 1996). The establishment of fully interoperable radio networks will require multimode, multiband communications capabilities, which are the focus of several DOD-funded research and demonstration programs.

The military ideal of a multimode, multiband radio implementing many different waveforms over a broad frequency domain does not have a commercial counterpart, although commercial technology that would support multiple standards is being explored under the European RACE and ACTS efforts. Existing multimode commercial systems have at most six different types of waveforms, each one generally restricted to a narrow frequency range, sufficient to access all of the wireless services that an international traveler is likely to need. The commercial sector is unlikely to support additional waveforms because of the costs involved, whereas the military would almost certainly pay for them to gain the added functional flexibility.

3.4.3.1 *Software-Defined Radio*

Software radios are evolving in both the defense and commercial sectors. The military version is intended to enable interoperability among defense networks, reduce logistics support costs, and provide the capability to add new functions to fielded equipment through software updates.[19] The commercial work is driven by the need to accommodate the large number of standards used in mobile telephony. The design of common hardware for a wide range of applications would offer convenience to consumers and simplify manufacturing; however, the ultimate popularity of these systems will depend on whether they prove to be cost competitive with multiple dedicated implementations.

Several DOD-funded experimental models have been built. In field demonstrations, SpeakEASY was shown to be capable of receiving communications from the Air Force and translating them for the receivers and networks used by Army ground forces. The four-channel radio is compatible with some legacy waveforms and spans frequencies from 2 MHz to 2 GHz. The ACE, JCIT, and Millennium programs are not yet completed. Software radios are also being designed under the GloMo program to have adaptive interference-rejection capabilities. It is not yet clear whether any of these systems will offer the performance and cost effectiveness needed to initiate a production program.

Most commercial dual-mode digital cellular and personal-communications units can implement multiple transmission and reception formats using DSP software. Information about commercial radios still in development is typically not publicly available. There are undoubtedly plans to make software radios, which will likely be less flexible than are military versions. The commercial radios may contain software that is not intended to be changed after manufacturing. Furthermore, they will likely not offer the frequency range, extent of waveform synthesis, or sophisticated security expected for military applications.

Meanwhile, the commercial sector has focused intensive R&D efforts on various radio components to achieve incremental, practical advances. The DOD can expect to take advantage of the rapid commercial progress in many components—A/D converters, DSP chips, RF amplifiers, display elements, processors, batteries, and storage devices—which will probably drop in price over the next several years. However, as discussed in Chapter 2 (Section 2.4), the DOD will likely need to develop its own specialized filters that can accommodate a broad range of frequencies and bandwidths, as well as antennas that offer both frequency and beam-shape agility.

When all the functions of a radio are defined by software, the "intelligence" and network services offered by the radio can be extended to greatly enhance military applications and perhaps eventually lead to intelligent radio services in commercial applications as well. Smart radios (i.e., radios capable of optimizing frequency, modulation, and protocols for a given purpose and signal environment) can incorporate the rules learned by an experienced communications specialist. Many simple rules define how to minimize interference. These rules can be applied in real-time, packet-based communications systems much more effectively than in traditional voice systems. Through real-time evaluation of each communication link and the spectrum in which the system operates, new levels of intelligence can be achieved to avoid jamming or to optimize transmissions under a wide variety of conditions (e.g., by minimizing battery drain, reducing traffic in the vicinity of hostile jamming activities, maximizing bandwidth or network capacity).

The introduction of the multimode software radio creates a significant opportunity for the convergence of many different systems and functions. Traditional defense platforms have separate systems for communication, navigation, identification, data exchange, signals intelligence, electronic warfare, and other functions. A software radio could be rapidly configured to perform any of these functions in any combination required. This convergence of technology will reduce the numbers of military systems procured while also increasing the cost effectiveness and utility of equipment. The resulting lightweight, agile platforms will be capable of rapid response to support the small units of fast-moving military forces now evolving. The increased availability, utility, and power of radio devices will create a new paradigm for military communications (see Table 3-2).

3.4.3.2 Co-Site Interference

Co-site interference, which is already a problem for military communications platforms, will worsen with the introduction of multimode, multiband radios unless new mitigation approaches are developed. Current technology designed to reduce the effects of co-site interference on radio performance is quite limited. Power combiners can connect up to five transmitters to a single antenna, but only if the frequencies are sufficiently separated. Receive co-site filters can suppress the carrier of co-located transmitters, but broadband signals are not suppressed adequately, and the broadband noise of transmit power amplifiers is not suppressed sufficiently at frequencies near the transmitting frequency.

TABLE 3-2 Current and Emerging Military Communications Paradigms

Current Paradigm	Emerging Paradigm
Radios are a precious resource. Hardware developed decades ago remains in low-rate production with high unit cost. Radios are rationed to one or fewer per platoon.	**Radios are a ubiquitous resource.** Like computing, communications will become so inexpensive that it will be widely available. As with computers, new radios are implemented as applications software running on standard platforms.
Legacy radios. Radios are standardized but there are too many standards, all based on legacy hardware. Few radios are compatible with other radios even within the U.S. inventory. The problem is amplified when the Allied and multinational radio systems are considered.	**Interoperability.** Radio waveforms, bandwidths, channels, modulations, error correction, and cryptography are all implemented in software. Even legacy systems are implemented in software, which might even be downloaded over the air.
Manual mode and frequency selection. Radio operators select frequencies and modes based on command instructions. Radio units are often manually turned off (e.g., "radio silence") and communications cease.	**Connectivity.** The radio decides which modulation, frequency, and power level is best based on "RF situation awareness" (the need to remain covert), the available resources, and the amount of data that need to be transmitted.
Communications van complex. All the various radios need to be housed in large communications vans, which are scarce resources.	**"Palm top" communications complex.** Multibands, multimodes, and multichannels are on a card; connectivity is achieved by "cleverness" and resource allocation, not brute-force transmit power and antennas.
Radio as a communicator. Each radio has one channel for voice or messaging.	**Radio as a sensor.** Each radio has many channels. Some can be programmed and used in networks for intelligence collection and emission location by triangulation.
Development lead time. Radio upgrades and developments are measured in years.	**Rapid prototyping and deployment.** New features and upgrades are implemented in software, with "few" hardware changes required.
Individual radio units. Communication is point to point on a single link that is limited by output power, sensitivity, and waveform propagation.	**Communications network.** The full set of protocols, including TCP/IP message routing, is implemented. Whole groups of units work together to achieve connectivity.
Closed, proprietary architecture. Fully functional "black box."	**Open architecture.** Standard interfaces and packaging are the norm, with plug-and-play hardware and software modules.

Moreover, receive co-site filters become complex when the number of co-site transmitters is three or more, and receiver noise performance is degraded, resulting in reduced transmission range and an increased error floor. Co-site problems extend to antenna beam shape, which changes when antennas are used in close proximity to each other or to metallic structures. Because of the unique conditions on military communications platforms, R&D in this area will likely need to be supported by the DOD.

3.5 DEFENSE TECHNOLOGY POLICY ISSUES

The government influences private-sector technology development in a variety of ways. The instruments of government policy include indirect methods, such as investment tax credits, or direct methods such as federal funding for R&D and technology testbeds. Sometimes these policies are implemented to accelerate the development of strategically important technologies; at other times the motive is to ensure that equipment will be available for procurement by the government in a timely fashion.

Government policies supporting the development of appropriate defense technologies have always been a special case. In the past, when defense requirements generally guided private-sector technology advances (e.g., transistorized components), federal investments in R&D were not controversial. Now that sophisticated consumer and industrial products are developed independent of defense requirements, the need for federal investments may seem less pressing. However, the DOD needs to maintain a competitive advantage over potential adversaries with respect to warfare capabilities, including communications systems. The technology policy issue for the future is how to encourage innovations in electronics and communications technology that will dominate world markets while also ensuring that the U.S. military retains capabilities that exceed those of potential adversaries.

3.5.1 Implications of Changes in Military Tactics

The Gulf War demonstrated the way in which high technology permeates warfare. Advanced sensing, imaging, and targeting capabilities in the Patriot missile defense system, stealth aircraft, and other systems provided extensive advantages for U.S. forces. For example, Patriot missiles were aimed using surveillance satellites controlled from the United States. Liftoffs from Iraq were observed by these satellites within seconds, and critical targeting information was relayed through controllers in Colorado to the front-line Patriot batteries. This orchestrated activity demonstrated the capabilities of the U.S. military's existing global communications network, which required the support of high-bandwidth data links to move sensor information both to and from the theater of action. But the Gulf War experience also suggests that communications advances are needed to enable rapid infrastructure deployment, logistics enhancements, and increased protection of technologies to prevent their exploitation by adversaries.

3.5.2 Rapid Infrastructure Deployment

During the ground war, the mobile forces moved so quickly that the communications infrastructure could not keep up with the front lines. Future communications systems will likely need to be rapidly deployable (and redeployable) so that they can keep pace with rapidly developing battles. Because Iraq did not react when U.S. troops first began arriving in Saudi Arabia, the coalition forces were able to build up an overwhelming combat strength in the Middle East as well as the logistical stockpile needed to pursue vigorous modern warfare. Adversaries in future wars are unlikely to be so accommodating, meaning that forces will need to be projected rapidly from the U.S mainland. Future conflicts are likely to be "come as you are," and communications infrastructures will need to support immediate action.

The recognition of this need has heightened interest in "instant infrastructures" based on satellite communications and mobile elements. The RAP has been proposed as a basis for a moveable front-line infrastructure with sophisticated, on-the-move antenna systems able to maintain high-bandwidth, point-to-point links with the rear-area infrastructure. To avoid the latencies inherent in satellite communications, hybrid systems that consist of DBS downlinks and UAV uplinks are being investigated. In general, these systems are viewed as backups to the terrestrial trunk linkages.

Continued military R&D investments will probably be needed because there seems to be little commercial interest in moveable infrastructures. One example of a commercial system with moveable elements is the Metricom multihop packet radio network, which operates in the unlicensed ISM bands in the San Francisco Bay and Washington, D.C., metropolitan areas. Although the infrastructure radios are in fixed locations, the multihop architecture makes it possible to add coverage in an incremental fashion through the addition of relay radios within the service area; bandwidth can be added also.

3.5.3 Logistics

Future military communications systems will need new features corresponding to the reduced size of U.S. forces. Current planning provides forces that are only sufficient to fight two regional conflicts at the same time. Instead of stationing so many troops overseas in areas of high tension, a split-base approach will be used, with advanced echelons overseas and the bulk of the forces on the U.S mainland. This approach will require high-quality, high-bandwidth connectivity worldwide, complete with access extensions that can be rapidly deployed, torn down, and reestablished as troops move.

Logistics tracking and management will be especially critical, given the growing need to transport materiel from the United States to the scene of the conflict. Many commercial systems are available. For example, OmniTRACS makes it possible to track vehicles continuously as they move and to plan routes efficiently. Package delivery services such as UPS and Federal Express have deployed sophisticated logistics systems for tagging packages and tracking them en route while also providing user-friendly on-line services that enable shippers to find their shipments. Wireless LANs were originally developed partly for warehousing applications. Finally, wireless tagging technology could provide the DOD with automatic inventory and location-identification capabilities, providing the basis for a complete logistical information system that could track the location of every item shipped.

3.5.4 Preparing for Unsophisticated Adversaries

There is some uncertainty about the technical requirements for communications during future confrontations with unsophisticated adversaries. Recent U.S. actions in Haiti and Somalia are examples of these types of operations, which may become more common as the United States plays an expanding role in peacekeeping and peacemaking missions. These countries tend to have little modern communications infrastructure, although this situation is changing as worldwide markets evolve for advanced technology. When deployed in less-developed countries, the U.S. military could bring along state-of-the-art commercial infrastructure technology. These systems would need to be shipped, installed, and operational within days, with military systems sufficing in the meantime. The commercial systems could transport the bulk of noncritical traffic, making it accessible to a smaller number of military-specific systems in the field.

In many ways, peacekeeping and other nontraditional military operations are similar to law enforcement activities, and many of the same communications issues need to be addressed. Even an unsophisticated adversary could disrupt service to U.S. forces using commercial systems. For example, cellular infrastructure is difficult to hide and could easily be targeted for sabotage. Although stealth and LPD are not always critical to defense communications, steps need to be taken to prevent adversaries from learning of upcoming operations, performing traffic analyses, and intercepting specific types of communications traffic. The basic security and authentication mechanisms in the latest commercial systems can reduce interception by the technically unsophisticated; they are sufficient for nontactical communications traffic such as logistics support. Military-specific systems will continue to be needed for transmissions that require complete security.

The DOD might need cooperation and technical information from U.S. or foreign manufacturers so as to monitor the traffic of adversaries, track specific telephones, or infiltrate existing communications systems in particular countries. The U.S. military therefore needs to maintain a technical awareness of foreign-made equipment, perhaps as part of the effort to demonstrate, test, and procure COTS wireless technology (see Sections 3.2 and 3.3).

3.5.5 Preparing for Sophisticated Adversaries

Sophisticated communications technology is rapidly becoming a commodity. During the Gulf War some military specifications and procurement procedures were abandoned in an effort to get new capabilities, such as GPS, into the hands of the troops. Any adversary could buy the same sophisticated technologies; the threat is measured by how much the adversary can afford. Indeed, one of the implications of the Gulf War as a model for future conflicts is that the United States might not prepare sufficiently to recognize or defend against sophisticated adversaries.

A sophisticated adversary can be defined as one with the technical capability to build advanced communications systems or the financial resources to purchase what it needs on the global arms market. The greatest immediate threats are countries that can buy technologies from the countries that make them; for instance, the SCUD missiles used by Iraq in the Gulf War were based on the Chinese Silkworm missile.

To maintain a competitive advantage against these adversaries, the U.S. military could add military-specific modifications, such as security or waveform hiding, on top of commercial core systems. The military can leverage many commercial technologies, among them advanced ICs, DSP chips, and protocols. The advantage gained will depend on how these capabilities are integrated into defense systems and the choice and performance of the added military-specific capabilities.

3.6 SUMMARY

The DOD has many reasons to use commercial communications products and practices whenever possible, building on a long tradition of synergy between the two sectors. Many COTS technologies offer cost and performance advantages, and their quality is better than ever. The economies of scale achieved in mass production provide additional benefits and lessons that can also be exploited by the military. The selective use of commercial products and practices in DOD systems could help accommodate growing needs for global, untethered communications systems in spite of declining defense budgets.

However, the military will continue to have some unique needs that cannot be met by consumer products, or even future commercial R&D programs, because the motivations and interests of the two sectors differ. The DOD has unusual needs in three fundamental areas: network architecture, which influences all other aspects of a communications system; security, which encompasses confidentiality, data and system integrity, and service availability; and multimode, multiband systems, which can enable interoperability among diverse systems. The DOD needs to examine its needs in these areas carefully and probably pursue its own R&D in selected technology areas. All of these issues are addressed further in Chapter 4.

NOTES

1. For example, advanced coding (Cacciamani, 1970, 1971, 1973) has been used in commercial satellite communications since the early 1970s for both data and highly compressed digital imaging, enabling the use of antennas on the order of 18 inches in diameter for digitally compressed video signals with link BERs less than 10^{-9}. The best known of these technologies is probably CDMA, which has been widely adopted for cellular and personal communications systems worldwide. Encryption, along with data mining and RF fingerprinting, is increasingly being used to protect against fraudulent use in cellular systems, video entertainment subscription receivers, and business data. Finally, on-board digital processing will be used in the planned mobile telecommunications satellite and high-speed data satellites such as Teledesic.

2. A short lead time in a growing market can result in a large increase in market share. In addition, because prices can fall quickly after a new product is introduced, the first to market is often the only competitor to make a substantial profit. Yet a release date is often difficult to predict. Companies can be punished by the market if they fail to meet predicted release dates, as often happens, for example, with software upgrades.

3. The internal design cycle may actually be much longer because the basic equipment architecture is more likely to be on a two-year design cycle paced by the evolution of new semiconductor components. During the baseline design cycle of up to four years, anywhere from one to four design teams may be working on the next baseline architecture.

4. Many U.S. commercial wireless communications suppliers include divisions that have historically been involved in defense work. Within these compa-

nies, cross-fertilization between the defense-related and commercial units may provide a mechanism for meeting military surge needs using the company's commercial products. However, this type of crossover is not always straightforward because of the differences between defense and commercial markets.

5. For example, current regulations regarding processors, A/D converters, and cryptography appear to reflect technologies that are nearly a decade old. The advent of common high-performance microprocessors enables the widespread development and use of cryptographic algorithms, which are often distributed on the Internet. The export of A/D converters is limited to technology of less than 8 bits, but advanced sigma-delta technology has only 1 bit (noise shaping and DSP techniques are used to increase dynamic range). Thus, the number of bits no longer seems like a useful metric for A/D converters; the metrics used to evaluate microprocessors seem equally outdated.

6. This is a simplified description of the decision-making process. More precisely, throughout the design, fabrication, and deployment of commercial products, trade-offs are made among performance requirements, standards requirements, cost goals, and design approaches to define a product that would be the most attractive and competitive in the marketplace. International, national, and regional standards determine many commercial design parameters, including off-axis emission from an antenna, maximum power flux radiated to Earth from a satellite, the capability of system users to coordinate or coexist with other users of a frequency band in the same geographic location, and numerous electrical safety regulations (e.g., related to wiring, batteries, radiation hazards, and chemical exposure).

7. Customers understand and expect this and are generally not willing to pay for a capacity that sits idle most of the time. Even during the busiest hour of the average business day—conditions that the systems are engineered to handle—there is a measurable probability of blockage that is calculated based on customer willingness to pay. Because the cost of a blocked call is usually only the effort required to try again shortly, there is little incentive to reduce the probability of blockage to zero. An interesting demonstration of the customer's acceptance of blockage and delay is the phenomenal growth of the Internet, where service is provided on a best-effort rather than guaranteed basis (although data services continue to come under increasing pressure for better service access).

8. Intel Corp., which after marketing its Pentium microprocessor found a design flaw in the precision of certain mathematical operations, uses a test suite comprising of billions of instructions to validate each possible instruction, register, arithmetic function, interrupt process, and instruction trap as well as sequences of events to prevent any surprises in complex applications. Only now are academic researchers considering more sophisticated theoretical techniques for dealing with testing processes of such enormous complexity. This research is critical to the future success of complex systems.

9. For example, commercial processes might take place at temperatures ranging from 0 to 50 degrees Celsius (°C) rather than –55 to 125 °C as in military processes. Or, commercial processes might involve 30 *G* of force rather than 1,000 *G*.

10. The NES is an encryption system certified by the National Security Agency

that enables clusters of defense computer networks to interconnect through the unclassified Internet. The NES provides high levels of assurance that a system communicates only with other systems that have comparable security levels.

11. Several military initiatives, including the Multilevel Information Systems Security Initiative and the DOD Goal Security Architecture, are intended to deal with various aspects of infrastructure in an effort to enable interoperability among systems. However, these programs have yet to field functions that enable communication between independent defense networks.

12. The ACN is designed to provide hierarchical communications over a broad theater of operations. Cross-linking and networking will enable various networks to communicate and access services through satellite links worldwide. The ACN will also serve as a repeater, picking up signals and rebroadcasting them over and around terrain obstacles, thereby extending the range of low-power equipment used on the ground.

13. In a base-station-oriented architecture, a greater investment is ordinarily made in the base station than in terminals. In such a network, both the transmit and receive link equations can benefit from the improved performance of larger antennas, more powerful transmitters, and more sensitive receivers. In typical systems the link advantage relative to the peer-to-peer design is approximately 10 dB.

14. Security is an issue in equipment deployment: The use of systems with cryptographic security requires procedures for securing clearances and equipment controls.

15. The TCP/IP protocol suite would need to be supported on top of ATM because the DOD has identified TCP/IP as the means for ensuring interoperability across heterogeneous military networks and because the entire system is unlikely to be constructed from native ATM technology.

16. An additional drawback is ATM's strong connection orientation, which makes it difficult to support mobility because existing connections need to be broken and reconstructed repeatedly. Furthermore, the ATM cell (i.e., data packet) structure was designed for the extremely low BERs of fiber-optic communications, whereas a radio fade can persist for several cell durations, making it difficult to use standard coding techniques to improve link quality. The loss of even a small number of ATM cells in a highly stressed network can dramatically reduce packet throughput.

17. The importance of distilling source information prior to transmission over a network is well understood in the commercial sector but remains an issue for the military, especially the Army, where communications, command-and-control, and intelligence functions are separate. There is no financial incentive on the part of the command-and-control and intelligence communities to spend resources to distill data at the source. Often the problem is passed off to the communications community, which is forced to transmit whatever is provided. For example, in situation awareness (SA) reports, positions are reported every 12 seconds regardless of motion. As a result the communications system is overloaded with SA reports. A more efficient approach would be to project positions based on direction and velocity and only send reports when the trajectory or velocity changes. But such an approach would require the development of software at a cost to the command-

and-control community. Instead the practice has been to blame the communications system for failing to support the traffic load. This situation would never arise in the commercial cellular industry, where providers take a systems approach and make trade-offs between bandwidth costs and source compression costs.

18. Mobile code, such as Java, might eliminate the need to agree on a compression standard because the delivery of executable code (along with the transmitted data) would allow the receiver to adapt to the sender's coding scheme.

19. An alternative approach would be to implement new functions in ASIC chips, which offer efficiencies in terms of power consumption. However, this approach would not provide an open architecture and might not be adaptable to future radio waveforms.

4

Conclusions and Recommendations

The history and challenges of wireless communications, as outlined in previous chapters, suggest a variety of strategies that could be pursued to fulfill the vision for untethered military communications systems. This chapter summarizes and integrates key points made in the preceding chapters to provide a set of 12 recommendations directed to the DOD and DARPA. Organizational changes are recommended that would provide an environment conducive to the development and military application of state-of-the-art commercial technology. To meet defense-unique needs, specialized R&D and demonstration efforts are recommended that focus on various aspects of wireless technology, from the highest network level down to individual components.

4.1 HISTORY AND CHALLENGES OF WIRELESS COMMUNICATIONS

Voracious consumer demand is stimulating many advances in commercial wireless communications technology, particularly cellular and cordless telephones. The portfolio of wireless services now available in the commercial marketplace includes a wide range of telephony, paging, and data applications delivered over a variety of service offerings ranging from land mobile radio to cellular to satellite communications. Each service offers a unique combination of coverage region, bandwidth, subscriber equipment properties, and connectivity. In the aggregate, commercial wireless capabilities are considerable, yet many technical challenges remain. The cost of wireless voice systems needs to be reduced and their quality improved. Specialized wireless data networks have not taken off as yet, perhaps because they are not powerful enough or because mass market applications have yet to emerge. Considerable research to address these and other issues is under way, both in the United States and overseas. Industry road maps suggest that, by early in the twenty-first century, commercial wireless communications will meet the long-term goal of enabling users to communicate "anytime, anywhere."

The DOD uses a variety of wireless systems that are based on 1970s and 1980s technology and designed to serve specific needs. The DOD no longer drives the evolution of state-of-the-art communications technology but still needs access to it, perhaps more than ever. As threats to peace change from global to regional conflict, a transformation is taking place in military roles, missions, and communications needs. The vision for military communications stresses C^4I and the protection of the lives of U.S. personnel, who will be based principally in the United States but will need to be prepared to move quickly throughout the world to carry out a variety of missions, including noncombat roles such as peacekeeping and humanitarian response. Such missions are nontraditional in the sense that coordination with foreign partners may be essential, whereas national survival will not be at risk as was anticipated during the Cold War. In addition, the need for U.S.-based logistical support will grow, and new systems will be required to counter terrorism. Thus the accurate, timely transmission of information will be perhaps more essential than ever in meeting military objectives. Effective global communications systems will be critical.

The civilian and military sectors have a long history of interaction in the design and deployment of wireless communications technology. In the Gulf War, DOD used commercial wireless equipment such as GPS receivers and found that the performance was comparable to that of equipment designed explicitly to meet military needs. Yet current commercial technologies and practices cannot meet all military needs. For example, the military cannot tolerate the long lead times—on the order of months to years—that are typical in the building of commercial communications infrastructures. Commercial wireless companies carry out elaborate advance planning and measurement operations, whereas the military requires networks that can be organized quickly and can adapt rapidly to changing operating conditions (including spectrum availability). These networks will also have to be compatible with other military communications systems, both new and old.

Differences between military and commercial needs also have implications for network architecture. Commercial research on integrated (i.e., multimedia) systems is oriented toward network architectures based on the base-station-oriented model. It is not clear whether that approach or the peer-to-peer design will be more appropriate in future military settings. Previous DARPA research on packet radio networks has encompassed base-station-oriented, peer-to-peer, and multihop networks.

The evolution of technology is also influenced by organizational differences between the two sectors that encompass market drivers, acquisition practices, and maintenance and repair arrangements. Commercial R&D and manufacturing are oriented toward mass markets of tens of millions of units. Adding functionality to designs has to be justified by consumer demand. By contrast, military equipment is designed to provide the functions required to fulfill missions. The number of units produced is tiny relative to commercial markets. Commercial customers acquire equipment and subscribe to services as they are ready to use them, expect effective function within minutes, and rely on equipment manufacturers for repairs. The military acquires equipment on a contingency basis, operates its own repair facilities, and is prepared to train its personnel to use communications systems.

Many research efforts are under way to realize the commercial and military visions for wireless communications. Fueled by the success of cellular communications and projections of ever-expanding markets for wireless services, the commercial sector is pushing ahead in various areas. One objective is to enable portable devices to communicate at the high bit rates needed for advanced information services. Another objective is to advance the state of the art for software radios as a means of fostering economies of scale in R&D and manufacturing in a world of diverse and changing technical standards. By using multiple types of operating software, such radios can serve as single hardware platforms capable of transmitting and receiving signals that conform to a variety of standards. Meanwhile, DOD is taking a dual approach to wireless technology development by both conducting its own research, focusing primarily on components, while also relying increasingly on commercial technologies to ensure interoperability and systems integration. The DARPA GloMo program has initiated a broad range of coordinated R&D efforts that will provide enabling technologies for future military systems. Among the commercial technologies that the military expects to use are Internet protocols and ATM.

These observations lead to the following general conclusions:

- A large gap remains between public expectations for mobile communications ("anytime, anywhere") and the available technology.
- Over the next 10 years or so, market forces will fill this gap by developing new technologies for commercial wireless communications.
- The military thus has much to gain from positioning itself to use COTS communications equipment to the greatest extent possible.

- Some military needs for wireless communications technologies will exceed or differ significantly from anticipated commercial developments. For example, the military has unique concerns with respect to network design, security, interoperability, and multimode/multiband systems.
- The commercial sector has its own incentives to produce advanced communications devices, components, and subsystems as well as complete systems. To use commercial technologies effectively, the DOD will have to take special measures to promote the development and acquisition of COTS products that can be integrated into systems that meet specialized military requirements.

These conclusions provide the basis for the recommendations presented in the remainder of this chapter. The 12 recommendations are organized in a hierarchical order from the general to the specific. The first three focus on organizational changes, or meta-issues, that need to be addressed by DOD to help align military requirements with commercial products and services and provide an environment conducive to the absorption of state-of-the-art technologies. The other nine recommendations identify R&D projects that should be carried out by DARPA to advance the synergy between military and commercial systems while also meeting specialized military needs that will exceed anticipated commercial developments. The R&D recommendations are presented in order of priority, reflecting the committee's view that high-level systems issues are of paramount importance. The recommended research excludes subjects that will be adequately covered by the commercial sector.

4.2 STANDARDS DEVELOPMENT

1. The DOD should participate in standards-setting activities for wireless communications technologies and systems.

With commercial demand for wireless technology growing worldwide and DOD budgets flat or falling, incentives for future commercial-military synergy need to be provided by the military side. Consequently the defense community needs to gain a deep understanding of technology trends so as to obtain advance notice of new concepts and influence the development of cost-effective equipment that meets military needs. Although new technologies can originate in diverse settings that include industry, academia, and nonmilitary government laboratories, the features of available equipment are determined to a large extent in the process of standards setting.

Current R&D and standards activities are focused on enhancing wireless communications technology to make it possible to move many types of information (including data, video, and images) to and from portable wireless devices. Although this work is certain to produce new technology, the commercial deployment of these innovations is not assured. The availability of innovative technology in the marketplace depends on business, social, and government policy factors. Military planners need to maintain a continuing awareness of the difference between what is possible technically and what is available in the market to meet military needs.

Because military requirements often exceed those of the commercial market, opportunities for the DOD to use commercial products depend on equipment details, which are reflected in standards. Such details might be of little initial interest to consumers. Yet the defense community is often ahead of industry in recognizing features that will eventually be important to all users; if these needs are not addressed in the standards-setting processes then they might have to be met later in commercial settings at great cost, both financially and in terms of network performance. An example is the poor state of network security in analog cellular systems, an issue not considered in the initial design process. Newer digital systems help alleviate this problem, but adding security features to older equipment remains an expensive and difficult challenge. The military, by contrast, always plans for system security in advance.

Standards also influence whether "hooks" or interfaces are designed into commercial technologies to enable modifications that would meet specialized military needs. By participating in the standards process, government agencies will make it possible to embed standard devices in military-specific system architectures and generally promote a capability for cost-effective systems integration.

The DOD could influence standards setting by participating in activities such as the ATM Forum, the IETF, and the Multimode Multiband Information Transfer System Forum. Effective participation will be constrained by the rules governing standards organizations and by the abilities of DOD management and participating individuals to influence technical decisions and political processes. They will need to understand how standards for wireless communications are established, a complex process that has influenced the very different evolutionary paths of wireless technologies in the United States, Europe, and Japan. In addition, the DOD could benefit from the analysis, simulation, and laboratory testing of candidate technologies to support defense interests in the establishment of specific standards. These activities could reveal the limits of some technologies that would not be apparent in more benign commercial settings but are likely to be of long-term significance to consumers. By conducting such tests and sharing the results with the commercial sector, defense agencies will improve the chances that adopted standards will serve immediate military needs and also benefit the commercial sector in the long run. Participation in standards creation could ultimately prove to be more cost-effective than commissioning equipment explicitly designed to meet military needs.

4.3 DEMONSTRATION AND TESTING RATHER THAN DEVELOPMENT

2. The DOD should pursue a vigorous process of technology demonstration and testing prior to development and procurement. In particular, the focus should be on system concepts based on commercial technologies and specialized military enhancements.

As the defense budget is reduced, less money will be available for major procurements. Nevertheless, an inventory of advanced technologies can be maintained and deployed as needed if military equipment is adapted whenever possible from commercial technologies. Examples of this strategy include the Condor project, in which DOD is supporting the development of a cryptography module on top of the cellular telephony system, and the Army's plans for the digitized battlefield. The Condor project is focusing on core noncommercial technologies such as on-the-move, high-bandwidth, phased-array antenna technology. The Army's digitized battlefield network will include switches, routers, and hubs based on commercially available technology. The Army is not building its own ATM switches, which are widely available in the private sector, but instead is developing high-speed encryptors and decryptors that are compatible with commercial switches.

Military planners need to understand the thrust of commercial developments and identify critical technologies that are unlikely to emerge—soon enough or ever—from the commercial sector. These technologies need to be developed by the DOD. For all other elements of wireless systems, new commercial developments need to be tested and evaluated on a continuing basis in military exercises to determine and verify their suitability for military use. Resources available to support this approach include the federal defense laboratories, which can provide a critical bridge between the military's operational needs and private technology development, and the defense industry, which is well equipped to collaborate in the development of prototype military-specific equipment. In addition, both the DOD and industry will continue to rely heavily on U.S. colleges, universities, and technical schools to provide competent engineers, technicians, programmers, operations staff, and scientists as well as key technological breakthroughs and innovative ideas vital to U.S. military and commercial competitiveness.

The DOD's technology demonstration and testing efforts need to take into account the unique communications needs of each military service. A mechanism is needed to help integrate the requirements of these varying systems and find a common architecture that would cost-effectively support the majority of those requirements. This task might be carried out by an R&D subunit such as DARPA or an existing Pentagon-level joint program office. In the past the military has focused on one radio subsystem at a time, with the result that interoperability has been minimal. The vision of the future calls for the development of an overall system concept with a standard set of multimedia protocols and waveforms to assure interoperability at all echelons, with individual radios specified and procured to work within this architecture.

4.4 PROCUREMENT

3. The DOD should plan a new approach to procurement that will identify how commercial infrastructure systems and subscriber equipment can best be used for military purposes and how to purchase commercial equipment in the most productive way.

The DOD needs to develop models to analyze how best to use commercial systems and equipment. For example, a planning system could be created to keep track of commercial communications infrastructure and service access points within a given geographical region. Such a system needs to be comprehensive, monitoring telephony, data, microwave, satellite, and fiber-optic services. The analysis could suggest the most effective means of information delivery for each source-destination pair, identifying the gateways and networking software required for interoperability and backup paths. This information could be used for planning or real-time robust network management, performance optimization, and repair.

A system is also needed to evaluate the potential for defense equipment based on commercial technology by comparing the cost-effectiveness and performance of legacy hardware to those of COTS products. In many cases COTS components may represent improvements for the military in terms of cost-effectiveness, power effectiveness, or other important features. The planning system needs to be capable of determining how to apply these internal components to anticipated defense applications. As part of this process the DOD needs to develop the expertise necessary to translate its operational needs into requirements for commercially available equipment.

Once the DOD identifies how best to use commercial technologies, cost-efficient acquisition strategies need to be pursued. Government-wide purchases might achieve greater economies of scale than would purchases by individual military services. The DOD could use this opportunity to become a more effective customer. Government acquisitions often add specialized requirements that prohibit commercial suppliers from offering bulk rates. As an alternative, DOD could explore how to purchase bulk quantities of standardized equipment and then separately acquire enhancements that align a system with military requirements. It might also be possible for designated agencies to acquire equipment and then distribute it to other government users. For example, the Army Signal Corps could acquire all DOD communications equipment and provide for all maintenance and upgrades on the basis of annual agreements. Consolidated purchasing might enable cost-effective reliance on commercial providers for maintenance, logistics, and training.

The DOD can also foster commercial-defense synergy by allowing multiple vendors to build subsystems that meet open hardware and software interface standards and by selecting, from the field of possible vendors, those most capable of creating an ongoing competitive production over a long, sustained product life. When procurement programs keep at least three vendors competitive over the course of the system, the competition encourages the evolution of advanced features and improvements in cost-effectiveness. The STU III is an example of a fully competitive acquisition in which three vendors competed for production of several hundred thousand secure telephones. In this controlled market, each vendor was required to meet open interoperability standards, but each was also allowed to implement unique features and functions to attract market share and compete on price. Each competitor's model updates kept the other competitors busy matching features and prices, a process that benefited both the users and the government while also motivating ongoing technology insertion. The SpeakEASY radio is another example of open system architecture. An "open system forum" allows contractors to participate in setting hardware and software standards that specify open interfaces between system components.

4.5 MODELING AND SIMULATION

4. DARPA should build on current research in modeling and simulation to incorporate the communications traffic, mobility of network elements, and radio propagation encountered in mobile military information networks.

The performance of wireless communications systems depends on three phenomena: communications traffic, mobility of network elements, and radio propagation. Accurate assumptions about these phenomena need to be made if the DOD is to design, deploy, and operate effective systems. Modeling and simulation are the best available tools for optimizing system design and predicting performance.

Various commercial packages are available for simulating communications systems consisting of established components and subsystems. The quality of the results they produce depends on the accuracy of the models of traffic, mobility, and radio propagation that they incorporate. Good models of narrowband radio propagation are available. However, models that incorporate site-specific characteristics of buildings, terrain, and foliage are not commercially mature, and there are no radio propagation models that incorporate adaptive antennas. Furthermore, available "teletraffic" models apply to only a few simplified conditions, and mobility models are confined to abstract formulations (e.g., fluid flow, Brownian motion) that have not been validated with reference to practical conditions. As a consequence, existing tools cannot provide realistic analyses of complex DOD systems. Much more is needed, especially in the site-specific channel modeling, traffic, and mobility areas, to provide military planners with the tools necessary to rapidly deploy wireless systems in battlefield scenarios in a wide range of modern theaters (e.g., built-up urban areas). DARPA should stimulate research to derive and validate models of wireless channel effects using modern modem and antenna technology so that appropriate protocols, applications program interfaces, and optimization algorithms can be developed.

Research leading to techniques for accelerating the run-time of complex wireless system simulations would benefit the military and commercial sectors alike. Simulation of communications systems is often performed by signal-processing workstation tools, which enable programmers to define a complex communication system by connecting icons representing predefined building blocks. This approach enables programming to be automated, but the resulting simulation code is often inefficient to run. DARPA research in this area could build on the impressive progress already made in the S3 program using parallel processors to simulate communications systems. This program includes a wireless component that could be enhanced to simulate the traffic, mobility, and radio propagation conditions of military communications.

DARPA could also establish libraries of code representing communications waveforms, protocols, source coding techniques, and network interfaces. This research would accelerate system design while also reducing the expense of simulation software maintenance, which typically costs 10 times more than software creation because of the longer time frame involved and rapid advances in technology. Research on system development and maintenance methods will help identify the most effective strategy.

Finally, integrated tools are needed to assess the performance of the subsystem elements of a software radio operating within a large-scale network in motion over a broad geographic area. These tools would enable researchers to explore network quality, connectivity, stability with respect to multiple performance criteria, and unexpected problems. Some research groups are developing their own tools to investigate specific aspects of performance. DARPA could attempt to provide an integrated capability built on a common application interface that would reveal the effects of each system component on the performance of the entire system. This capability would enable the optimization of an entire system rather than just the individual components.

4.6 NETWORK ARCHITECTURE

5. DARPA should initiate research to produce network architectures that incorporate commercial products in a manner that meets military requirements.

The performance of a wireless communications system depends in large part on the coordination of network elements. The architecture of a network defines these elements, identifies pairs of network elements that communicate directly, and specifies the protocols for that communication. Architectures for wireless systems differ according to how terminal modems are connected (peer-to-peer versus base-station-oriented design), how infrastructure elements are connected (hierarchical versus distributed), and the nature of communications with other networks. Given the special needs of military wireless communications networks, designers need to adopt a system architecture that takes maximum advantage of the capabilities of commercial devices and subsystems yet also provides unique interfaces and protocols as dictated by military needs. DARPA research in this area is likely to reveal new network architectures that use commercial products and services in an innovative way to serve military aims and make it possible to gracefully absorb new technology at the subsystem and component level.

Within a new network architecture, many key issues remain to be resolved with respect to protocols for wideband data services. Current commercial systems treat each application (e.g., Internet, telephony, video dial tone) separately instead of taking an integrated approach. Recent discussions of nomadicity underscore the importance of considering wireless and mobile components and conditions as part of heterogeneous and interoperating network contexts. Military wideband packet radios, which are designed from the outset for integrated information services, could provide a useful model for future commercial developments. However, research is needed to determine network topologies and protocols that make full use of the capabilities of advanced radios.

In creating a new network architecture, protocols and algorithms need to be optimized to meet the objectives of military operations. This is a complex task. As an example of the complex interdependence of algorithms and architecture, the design of effective routing techniques needs to aim for the following objectives:

- Routing to maximize data rate, with minimum transport delay and minimum delay variance;
- LPD/AJ routing around vulnerable spots;
- Routing for priority assignment and allocation of resources;
- Routing for battery power conservation;
- Routing for congestion avoidance;
- Routing for access to servers for specific real-time applications; and
- Routing to avoid anticipated propagation anomalies.

The difficulty of achieving all these objectives is compounded by the need to run several different applications on mobile terminals at any given time. Therefore, the routing protocol needs to process each packet individually to meet the QoS objective of the transmitting application.

In a mobile wireless network it should not be necessary for every terminal to assume the size, weight, power, or cost burdens associated with critical network services such as database servers, image servers, speech recognition and synthesis, transcoding, transcrypting, position location, and health and safety reporting. Research on network architecture can identify the best way to distribute services among network elements with the aim of minimizing the weight and power consumption of devices carried by personnel.

4.7 NETWORK SECURITY

6. DARPA should conduct research aimed at understanding and bridging the differences between security needs in commercial and military networks.

The commercial sector is developing a variety of data networking products and services that are likely to be integrated into single-hop base-station-oriented architectures. However, as noted in Chapter 3, the requirements for security in commercial networks are likely to be different from those in defense applications. For example, troops on the ground or ships at sea that do not wish to disclose their location require high

LPD/I performance levels, a requirement not faced by most commercial users. These differences need to be understood and accommodated if the military is to make use of commercial wireless technologies.

Research is needed to improve the information-processing features of encryption techniques. For example, the end-to-end encryption of all traffic would obviate the need for interfaces or bridges between interacting systems. However, current end-to-end encryption algorithms require considerable network overhead to establish the cryptographic synchronization. DARPA could seek improved methods of end-to-end cryptographic synchronization through multiple networks with lower overhead than is currently possible. With such methods, the time required to establish cryptographic synchronization between active network members would be reduced significantly.

DARPA also needs to design network protocols that allow for commonality of hardware and software interfaces as well as security differences that meet the needs of all applications. These protocols need to provide multilevel security, long identified as important for defense communications. The concept entails the shared use of a network by individuals with differing authorization to access information of differing levels of sensitivity. Multilevel security implies the management of access to computing and communications systems and to information transmitted or stored in those systems commensurate with individuals' authorization. In general, the design and implementation of multilevel security systems have been imperfect, and wireless and mobile applications compound the challenge.

4.8 HIGH-DENSITY COMMUNICATIONS PLATFORMS

7. DARPA should conduct research aimed at reducing co-site interference.

Military ships, combat aircraft, UAVs, and mobile battlefield systems all operate a variety of communications systems in close proximity to one another. The placement of numerous radios in the same general location typically results in interference and many compromises. Current technology designed to reduce the effects of co-site interference on radio performance is quite limited. For example, power combiners can connect up to five transmitters to a single antenna, but only if the frequencies are sufficiently separated. Receive co-site filters can suppress the carrier of co-located transmitters, but broadband signals are not suppressed adequately.

DARPA needs to catalog the co-site environment of military platforms, identify both common and unusual problems, and design more effective solutions that are useful over a broad spectrum, such as 2 MHz to 2 GHz. Specific co-site problems that require attention include the following:

- The performance of an antenna changes when it is used near another antenna or metallic structure, and it no longer provides the expected beam shape.
- Electronic equipment emits low levels of RF radiation from internal local oscillators and data buses—signals that represent interference to other radio receivers.
- Radio receivers transmit a small amount of the local oscillator frequency used to tune the radio, causing interference that is particularly problematic if the local oscillator is dithered, hopped, or modulated.
- Radios transmit signals not only on the intended carrier frequency and modulation but also (in an attenuated fashion) on carrier harmonics, intermodulation distortion products, intermediate frequencies, up-conversion local oscillator frequencies, and the broadband noise of each power amplifier stage.

This R&D would be timely because the full deployment of software radios operating in broad frequency ranges will depend on the development of techniques to overcome co-site interference and high-power in-band interference. When serving several subscribers in the same band and mode at the same time, SINCGARS radios would experience significant interference. Significant technical challenges are associated with co-site interference generated within the software radio itself. Additional issues arise in the effort to enhance mobility of forces. For example, consolidating multiple mechanized vehicles will produce a single command track, which now has six SINCGARS radios and the usual problems associated with multimode radios, regardless of whether the radio architecture is analog, digital, or software based.

Research is also needed to minimize the interference caused by the spurious emissions (spurs) and harmonics of co-site transmitters and receivers. When many radios are co-located or networked, it might be possible for each transmitter-receiver pair to agree on a common frequency plan to minimize spurs and harmonics or on common filtering plans to minimize co-site degradation. Additional research topics could include time and frequency management (to prioritize transmissions) and reception time slots (to minimize the impact of transmissions likely to cause interference to other receive channels). Such research would need to take into account the interleaving and coding present in each channel and the likelihood of interference. Finally, a new generation of highly linear filters is needed with improved agility and a wider range of operating frequency than is currently possible.

4.9 SOFTWARE RADIOS

8. DARPA should carry out research and demonstration projects designed to field software radio technology for military applications.

Software radio is a far more versatile technology than the term "radio" implies. These radios can operate as part of a network and perform a vast array of electronic and computational functions (e.g., database management, transcryption) through the downloading of software. A networked radio can function in ways not envisioned when the component is manufactured. In a military scenario, some radios can function as active interrogators while others act in a passive manner, undetected but coordinated by active units. Similarly, software radios can perform services that until now were unique to each application.

A software radio could be a leading part of the C^4I infrastructure: With appropriate software it could be applied to signal intelligence, electronic intelligence, communications navigation and identification, electronic warfare, information warfare, electronic countermeasures, missile tracking, guidance, or commercial paging or telephony. The cost-effectiveness of software radios increases with the number of available software functions, the ease of performing new tasks, and the ease of cooperating with other network systems to accomplish larger tasks. The multiple roles that can be played by software radios could have an impact on DOD's organizational structure, as services provided by individual organizations and procurements are combined in one system.

Commercial software radios are likely to have more limited capabilities (e.g., changing signals and bandwidth on a single frequency) than are military versions, which are being designed to span large frequency ranges and implement many legacy waveforms. However, the commercial sector is achieving rapid advances in many software-radio components, such as A/D converters, DSP chips, RF amplifiers, displays, batteries, and data-storage devices. The DOD can use these COTS products to good advantage. At the same time, DARPA needs to undertake specialized R&D focusing on antennas (Sections 4.10) and filters (Section 4.12) for military applications. Exploratory research on novel components and designs could also be beneficial (Section 4.13). In addition, to identify any necessary improvements and make optimal use of this promising technology, DARPA needs to demonstrate software radio technology on defense platforms where density, power, and weight are critical.

4.10 SMART ANTENNAS

9. DARPA should conduct the research needed to adapt smart antennas for mobile military applications.

Commercial applications for smart, adaptive antennas are limited to relatively low-cost and unsophisticated single-band units with limited flexibility in beam pattern. Moreover, virtually all existing adaptive antennas for mobile radio applications are designed for use at base stations rather than mobile units. Military applications require antenna functionality for several orders of magnitude of frequency coverage as well as electronic tuning, the coupling of more than one transmitted and received signal, and a diversity of beam shapes ranging from omnidirectional to pencil beams. Several technical challenges need to be overcome before such antennas can be produced. DARPA research needs to pursue the following objectives, among others:

- Achievement of useful antenna performance over a large frequency range;
- Development of low-cost techniques for implementing beam shaping with affordable companion electronics that implement the phase shifting and weighting (i.e., the electrical process of feeding antenna elements); and
- Improvements in the cost-effectiveness of frequency-agile couplers with built-in co-site filtering.

4.11 SMART WAVEFORMS

10. DARPA should conduct research to produce transmission techniques that adapt to a wide range of operating conditions.

Commercial communications systems tend to operate under predictable, stable conditions. Therefore, commercial waveforms—characterized by frequency band, bit rate, modulation method, and source coding and channel coding techniques—are designed for operation over a limited range of conditions. For example, high-tier systems are designed for one set of conditions, whereas low-tier systems are designed for a different set of operating conditions. The operating environments and transmission conditions encountered by military communications systems are more unpredictable and subject to change. Military systems would therefore benefit from transmission technologies that adapt to changing conditions.

Recent research has produced theoretical results on the optimization of bit rate, modulation method, source coding, and channel coding techniques. New DARPA research should be aimed at uniting the theory of adaptive modulation and coding with the emerging technologies of advanced software radios (Section 4.9), which promise a practical means of implementing adaptive schemes. The results of the recommended modeling and simulation research (Section 4.5) would provide valuable tools for performing this work. In addition to meeting military needs for adaptable systems, this work could serve the long-term needs of the commercial sector by providing core technologies for integrated personal communications systems that combine the advantages of several of today's separate systems.

One objective should be to develop modulation and coding strategies that allow the demodulator to acquire the properties of interference in the time domain, the frequency domain, or the constellation domain so that the interference can be avoided or canceled. These techniques would concentrate energy in the regions of frequency, time, or code space that provide the greatest link margin. Another objective should be to produce transmission techniques that establish a data rate consistent with interference and fading conditions. These techniques should allow for selection of a data rate over two or more decades of bit rate. They should also be designed for rapid carrier synchronization time, rapid timing synchronization, and minimum training time for learning channel conditions. The choice of data rate should be consistent with adaptive source coding, which gracefully degrades the perceived quality of the end-to-end application as the channel quality declines.

In addition, for defense applications new waveforms are needed that allow for very high spread-spectrum processing gain. Typical spread-spectrum techniques require considerable computational power to be detected by the desired receiver. The high signal-processing gain required to jointly discover frequency offset, baud boundary (or baud boundary offset), and spreading code alignment at these high processing gains is inconsistent with traditional portable radio design. In addition, new spread-spectrum waveforms are needed that allow for further reductions in the detectability of transmissions, direction, and properties of the spread signal.

4.12 FILTER TECHNOLOGY

11. DARPA should conduct research to overcome the limitations of current filter technology for use in military software radios and high-density platforms.

Both software radios and high-density platforms require advanced filters. Filters currently constitute 25 percent of the volume of a typical software radio and are used in receive preselectors, power amplifier output filters, local oscillators, and mixers. To a great extent the radio receiver's sensitivity and dynamic range are determined by the selectivity and the losses of the preselectors, and the radio's co-site performance is determined by the selectivity of output filters and the filters in the local oscillators and upconverters. Active and digital filters are not appropriate for these functions because they introduce noise that degrades performance. Handheld software radios would be improved by the miniaturization of filtering functions and improvements in frequency tuning range and selectivity.

There is also a need for filters that cover wide frequency ranges. Commercial radio equipment spans either a small frequency band or a few selectable bands. Military radios that span wide frequency ranges (such as 2 MHz to 2 GHz) require lumped filters built of inductors and capacitors to handle the low end of the frequency range and transmission-line techniques for filtering at the high end. Research is required to identify new circuit and materials technologies that allow for tunable, highly selective filters that span this entire range, operate at higher power levels, and take up less physical volume. In addition, new processing techniques are needed to enable the monolithic integration of highly selective filters with semiconductor devices and to produce multisection filters in more sophisticated shapes.

4.13 NOVEL COMPONENTS

12. DARPA should develop novel components to enhance the flexibility of software radios.

Software radios have been significantly enabled by novel components, notably DSPs and FPGAs, which allow new waveforms to be added to fielded systems through the installation of new software and hardware. Additional novel components could further enhance the flexibility of software radio architectures. For example, components could be designed to reduce the equipment size, weight, and power needed to accommodate military designs that incorporate a wide range of potential future waveforms and large numbers of legacy waveforms. Radio performance could be improved through research into new DSP architectures that have adaptive resolution, clock speed, instruction sets, memory architectures, and arithmetic functions designed to the specific signal processing of communications systems.

Similarly, novel FPGAs that either interconnect analog circuit elements or integrate analog and digital operation could be of great value. A chip containing analog circuit elements and FPGAs could be applied to the analog front-end functions of many communications systems while also providing a broad range of interfaces to other systems. Research is required to identify the semiconductor processes and circuit topologies that provide sufficient isolation between the various analog functions, interconnection with minimal loss, and circuits with sufficient versatility to be configured for many applications.

Finally, the monolithic integration of nearly the entire software radio function could provide experience in combining analog and digital signal functions on a common substrate. It is currently difficult to keep digital noise from coupling into analog circuitry and degrading performance. Research on how to implement wide-dynamic-range analog circuitry monolithically with digital circuitry would provide the basis for implementing an entire software radio on a single component. Such an effort needs to encompass the monolithic implementation of high-performance filters that are tunable over large frequency ranges.

Bibliography

Abramson, N. 1982. "Fundamentals of Packet Multiple Access for Satellite Networks," *IEEE Journal on Selected Areas in Communications* 10(2):309–316.

Abramson, N. 1985. "Development of the ALOHANET," invited paper, *IEEE Transactions on Information Theory* 31(2):119–123.

Abramson, N., ed. 1993. *Multiple Access Communications: Foundations for Emerging Technologies.* IEEE Press, Piscataway, New Jersey.

Abramson, N. 1994. "Multiple Access in Wireless Digital Networks," *IEEE Proceedings* 82(9):1360–1370.

Abramson, N., and E.R. Cacciamani, Jr. 1975. "Satellites: Not Just a Big Cable in the Sky," *IEEE Spectrum* 12(9):36–40.

Abrishamkar, F., and Z. Siveski. 1996. "PCS Global Mobile Satellites," *IEEE Communications Magazine* 34(9):132–136.

Acampora, A. 1996. "Wireless ATM: A Perspective on Issues and Prospects," *IEEE Personal Communications Magazine* 3(4):8–17.

Acampora, A., and M. Naghshineh. 1994. "Control and Quality of Service Provisioning in High Speed Microcellular Networks," *IEEE Personal Communications Magazine* 1(2):36–43.

Allard, K. 1996. *Information Operations in Bosnia: A Preliminary Assessment.* Strategic Forum 91. Institute for National Strategic Studies, National Defense University Press, Washington, D.C.

Anderlohr, G. 1969. "What Production Breaks Cost," *Industrial Engineering* 1(9):34–36.

Asta, D. 1991. *Recent Dynamic Range Characterization of Analog-to-Digital Converters for Spectral Analysis Applications.* Project Report AST-14. Lincoln Laboratory, Massachusetts Institute of Technology, Lexington, Massachusetts.

Ayonoglu, E., et al. 1995. "AIRMAIL: A Link Layer Protocol for Wireless Networks," *ACM Wireless Networks* 1(1):47–60.

Baier, P.W., et al. 1996. "Taking the Challenge of Multiple Access for Third Generation Cellular Mobile Radio Systems—A European View," *IEEE Communications Magazine* 34(2):82–89.

Balakrishnan, H., et al. 1995. "Reliable Transport and Handoff Protocols for Cellular Wireless Networks," *ACM Wireless Networks Journal* 1(3):469–482.

Balakrishnan, H., et al. 1996. "A Comparison of Mechanisms for Improving TCP Performance over Wireless Links," pp. 256-269 in *Proceedings of the ACM SIGCOMM Conference on Applications, Technologies, Architectures, and Protocols for Computer Communications.* ACM Press, New York.

Berrou, C., et al. 1993. "Near Shannon Limit Error-Correcting Coding and Decoding: Turbo-Codes," *Proceedings of the IEEE International Communication Conference* 44(10): 1261–1271.

Biggins, J. 1996. *The Two Headed Eagle.* St. Martin's Press, New York.

Booz-Allen & Hamilton, Inc. 1995. *Tactical Internet Systems Description for TFXXI.* Report for the U.S. Army Communications Electronics Command by Booz-Allen & Hamilton, Shrewsbury, New Jersey.

Bradley, G. 1996. "Military Maneuvers: Boeing Buying McD; Intel, Phillips Retreat," *Electronics News* 42(2148): 1, 36, 38.

Bray, J. 1995. *The Communications Miracle.* Plenum Publishing, New York.

Cacciamani, E.R. 1970. "Communications Satellites for the 1970's," pp. 559–572 in *Proceedings of the AIAA 3rd Communications Satellite Conference,* N. Feldman and C. Kelly, eds. Paper No. 70-420. Alpine Press Co., Los Angeles.

Cacciamani, E.R. 1971. "A Channel Unit for Digital Communications in the SPADE System," paper presented at the IEEE International Conference on Communications, Montreal, Canada, June 14–16.

Cacciamani, E.R. 1973. "Synchronization and Coding in Data Communications Satellite Networks," pp. 311–327 in *Proceedings of the NATO Advanced Study Institute on Computer Communications Networks,* R.L. Grimsdale and F.F. Kuo, eds. Noordhoff International Publishing, The Netherlands.

Caceras, R., and L. Iftode. 1995. "Improving the Performance of Reliable Transport Protocols in Mobile Computing Environments," *IEEE Journal on Selected Areas in Communications* 13(5):850–857.

Calhoun, G. 1988. *Digital Cellular Radio.* Artech House Publishers, Norwood, Massachusetts.

Calhoun, G. 1992. *Wireless Access and the Local Telephone Network.* Artech House Publishers, Norwood, Massachusetts.

Chuang, J.C.I. 1987. "The Effects of Time Delay Spread on Portable Radio Communications Channels with Digital Modulation," *IEEE Journal on Selected Areas in Communications* 46(2):375–380.

Computer Science and Telecommunications Board, National Research Council. 1994. *Realizing the Information Future: The Internet and Beyond.* National Academy Press, Washington, D.C.

Computer Science and Telecommunications Board, National Research Council. 1996. *The Unpredictable Certainty: Information Infrastructure Through 2000.* National Academy Press, Washington, D.C.

Couch, L.W. 1995. *Modern Communication Systems: Principles and Applications.* Prentice-Hall, Englewood Cliffs, New Jersey.

Cox, D.C. 1995. "Wireless Personal Communications: What Is It?" *IEEE Personal Communications Magazine* 2(2):20–35.

Cox, D.C. 1996. "Wireless Loops: What Are They?" *International Journal of Wireless Information Networks* 3(3):125–138.

Crawford, G.P., and T.G. Fiske. 1996. "Reflective Color LCDs Based on H-PDLC and PSCT Technologies," *Proceedings of the Society for Information Display International Symposium: Digest of Technical Papers* 27:99–102.

Crochierre, R., et al. 1983. *Multirate Digital Signal Processing.* Prentice-Hall, Englewood Cliffs, New Jersey.

Cross-Industry Working Team. 1995. *Nomadicity in the NII.* Corporation for National Research Initiatives, Reston, Virginia.

Davids, N. 1996a. "Personal Digital Assistants: Part 1," *IEEE Computer Magazine* 29(9):96–100.

Davids, N. 1996b. "Personal Digital Assistants: Part 2," *IEEE Computer Magazine* 29(11): 100–104.

Defense Advanced Research Projects Agency. 1996. Home page, http://www.ito.darpa.mil/ResearchAreas/Information_Survivability.html.

Economist, The. 1997. "Financial Indicators: Telecommunications," p. 119 in Vol. 342, No. 8007.

Evans, T.F. 1997. "Cellemetry-Telemetry via Cellular," pp. 1–11 in *Wireless Personal Communications: Advances in Coverage and Capacity,* J.H. Reed, T.S. Rappaport, and B.D. Woerner, eds. Kluwer Academic Publishers, Norwell, Massachusetts.

Forman, G.H., and J. Zahorjan. 1994. "The Challenges of Mobile Computing," *IEEE Computer Magazine* 27(4):38–47.

Gallager, R.G. 1985. "A Perspective on Multiaccess Channels," *IEEE Transactions on Information Theory* 31(2):124–142.

Garg, V.K., and J.E. Wilkes. 1996. Wireless and Personal Communications Systems. Prentice-Hall, Englewood Cliffs, New Jersey.

Gavish, B., and S. Sridhar. 1995a. "Economic Aspects of Configuring Cellular Networks," *Wireless Networks* 1(1):115–128.

Gavish B., and S. Sridhar. 1995b. "Models for Configuring Cellular Networks with Mobility," pp. 168–178 in *Proceedings of the 3rd International Conference on Telecommunication Systems.* ATSMA, Inc., Nashville, Tennessee.

Gerla, M., and J. Tsai. 1995. "Multicluster, Mobile, Multimedia Radio Network," *ACM Wireless Networks* 1(3):255–266.

Gilhousen, K.S., et al. 1991. "On the Capacity of a Cellular CDMA System," *IEEE Transactions on Vehicular Technology* 40(2):303–312.

Goldsmith, A., and S.G. Chua. In press. "Variable-Rate Variable-Power MQAM for Fading Channels," to appear in *IEEE Transactions on Communications* 2:815–819.

Goldsmith, A.J. 1997. "Capacity of Downlink Fading Channels with Variable Rate and Power," *IEEE Transactions on Vehicular Technology* 46(3):569–580.

Grant, M. 1996. "Signal Processing Hardware and Software," *IEEE Signal Processing Magazine* 13(1):86–88.

Gundmundson, B., et al. 1992. "A Comparison of CDMA and TDMA Systems," pp. 732–735 in *IEEE Vehicular Technology Conference Record.* Institute of Electrical and Electronics Engineers, Piscataway, New Jersey.

Hafner, K., and M. Lyon. 1996. *Where Wizards Stay Up Late: The Origins of the Internet.* Simon and Schuster, New York.

Headrick, D.R. 1991. *The Invisible Weapon: Telecommunications and International Politics 1851–1945.* Oxford University Press, New York.

Hill, C.H. 1997. "The Spoils of War," *The Wall Street Journal,* September 11, pp. r1, r4.

Holzman, G.J., and B. Pehrson. 1995. *The Early History of Data Networks.* IEEE Press, Los Alamitos, California.

INSTAT/SIA Information Services. 1997. "Semiconductor End Use Forecast." INSTAT/SIA Information Services, Scottsdale, Arizona.

International Telecommunications Union Task Group 8/1. 1996. Working documents from the 10th meeting, Mainz, Germany, April 15–26.

Jubin, J., and J.D. Tornow. 1987. "The DARPA Packet Radio Network Protocols," *Proceedings of the IEEE* 75(1):21–32.

Kahn, R.E., et al. 1978. "Advances in Packet Radio Technology," *Proceedings of the IEEE*, 66(11):1468–1496.

Katz, R.H. 1994. "Adaptation and Mobility in Wireless Information Systems," *IEEE Personal Communications Magazine* 1(1):6–17.

Katz, R.H., and E.A. Brewer. 1996. "A Case for Wireless Overlay Networks," paper presented at the 1996 SPIE Conference on Multimedia and Networking, MMCM '96, San Jose, California, January 27–February 2.

Katzela, I., and M. Naghshineh. 1996. "Channel Assignment Schemes for Cellular Mobile Telecommunication Systems: A Comprehensive Survey," *IEEE Personal Communications Magazine* 3(3):10–31.

Kelly, K. 1994. *Out of Control.* Addison-Wesley, Reading, Massachusetts.

Lauer, G.S. 1995. "Packet-Radio Networks," pp. 355–358 in *Routing in Communications Networks*, M.E. Steenstrup, ed. Prentice-Hall, Englewood Cliffs, New Jersey.

Leiner, B.M., et al. 1996. "Goals and Challenges of the DARPA GloMo Program," *IEEE Personal Communications Magazine* 3(6):34–43.

Leiner, B.M., et al. 1997. "Issues in Packet Radio Network Design," *Proceedings of the IEEE* 75(1):6–20.

Lewis, T. 1993. *Empire of the Air: The Men Who Made Radio.* Harper Perennial, New York.

Lin, S., and D.J. Costello. 1983. *Error Control Coding: Fundamentals and Applications.* Prentice-Hall, Englewood Cliffs, New Jersey.

Masini, G. 1996. *Marconi.* Marsilio Publishers, New York.

McCanne, S., and V. Jacobson. 1993. "The BSD Packet Filter: A New Architecture for User-Level Packet Capture," pp. 259–269 in *Proceedings of the 1993 Winter USENIX Technical Conference.* USENIX Associates, Berkeley, California.

Mooney, E.V. 1997. "More Consolidation Yet to Come in Shrinking Paging Industry?" *RCR*, May 12. Homepage, http://www.rcrnews.com.

Myles, A., et al. 1995. "A Mobile Host Protocol Supporting Route Optimization and Authentication," *IEEE Journal on Selected Areas in Communications* 13(5):839–849.

Naval Studies Board, National Research Council. 1997. *Technology for the United States Navy and Marine Corps, 2000–2035: Becoming a 21st-Century Force*, 9 Volumes, National Academy Press, Washington, D.C.

Nguyen, G.T., et al. 1996. "A Trace-Based Approach for Modeling Wireless Channel Behavior," paper presented at Winter Simulation Conference, Coronado, California, December 8–11.

Norberg, A.L., and J.E. O'Neill. 1992. *A History of the Information Processing Techniques Office of the Defense Advanced Research Projects Agency.* Charles Babbage Institute, Minneapolis, Minnesota.

Ogawa, K., et al. 1994. "Toward the Personal Communication Era," *International Journal of Wireless Information Networks* 1(1):17–27.

Padgett, J.E., et al. 1995. "Overview of Wireless Communications," *IEEE Communications Magazine* 33(1):28–41.

Pahlavan, K. 1995. "Wireless LANs," in *Personal Communication Systems and Technologies*, J. Gardiner and B. West, eds. Artech House Publishers, Norwood, Massachusetts.

Pahlavan, K., and A. Levesque. 1994. "Wireless Data Communication," invited paper, *IEEE Proceedings* 82(9):1398–1430.

Pahlavan, K., and A. Levesque. 1995. *Wireless Information Networks.* John Wiley and Sons, New York.

Pahlavan, K., et al. 1993. "Decision Feedback Equalization of the Indoor Radio Channel," *IEEE Transactions on Communications* 41(1):164–170.

Pahlavan, K., et al. 1995. "Trends in Local Wireless Networks," *IEEE Communications Magazine* 33(3):88–95.

Pahlavan, K., et al. 1996a. "Transmission Techniques for Wireless LANs," *IEEE Journal on Selected Areas in Communications,* special issue on wireless local communications 14(3): 477–491.

Pahlavan, K., et al. 1996b. "Trends in Local Wireless Networks," invited paper, *Proceedings of the IEEE* 33(3):88–95.

Peterson, L.L., and B.S. Davie. 1996. *Computer Networks: A Systems Approach.* Morgan Kaufmann, San Francisco.

Ramsdale, P.A. 1994. "Personal Communications in UK—Implementation of PCN Using DCS-1800," *International Journal of Wireless Information Networks* 1(1):29–36.

Rappaport, T.S. 1989. "Indoor Radio Communications for Factories of the Future," *IEEE Communications Magazine* 27(5):15–24.

Rappaport, T.S. 1996. *Wireless Communications—Principles and Practice.* Prentice-Hall, Englewood Cliffs, New Jersey.

Rhode, L., and T.T.N. Bucher. 1988. *Communications Receivers: Principles & Design.* Prentice-Hall, Englewood Cliffs, New Jersey.

Russell, J.E. 1994. "Universal Personal Communications: Emergence of a Paradigm Shift in the Communications Industry," *International Journal of Wireless Information Networks* 3(3):149–164.

Sari, H., et al. 1995. "Transmission Techniques for Digital Terrestial TV Broadcasting," *IEEE Communications Magazine* 33(2):100–109.

Sass, P. 1996. *Survivable Adaptive Systems ATD Final Technical Report.* U.S. Army Communications Electronics Command, Fort Monmouth, New Jersey.

Sass, P., and I. Eldridge. 1994. "Army Demonstrates Wideband on the Move Communications for Digitized Battlefields," SIGNAL 48(7):54-55.

Sass, P., and L. Gorr. 1995. "Communications for the Digitized Battlefield of the 21st Century," *IEEE Communications Magazine* 33(10):86-95.

Shannon, C.E. 1949. "Communication in the Presence of Noise," *Proceedings of the Institute of Radio Engineers* 37:10–21.

Shannon, C.E., and W. Weaver. 1949. *A Mathematical Theory of Communication.* University of Illinois Press, Urbana.

Sheng, S., et al. 1992. "A Portable Multimedia Terminal," *IEEE Communications Magazine* 30(12):64–75.

Spencer, N. 1988. "Comparison of State-of-the-Art Analog-to-Digital Converters," p. 75 in *Project Report AST-4.* Lincoln Laboratory, Massachusetts Institute of Technology, Lexington, Massachusetts.

Steer, D.G. 1994. "Coexistence and Access Etiquette in the United States Unlicensed PCS Band," *IEEE Personal Communications Magazine* 1(4):36–43.

Tanenbaum, A.S. 1996. *Computer Networks,* 3rd Ed. Prentice-Hall, Englewood Cliffs, New Jersey.

Taylor, M.S., et al. *Internetwork Mobility.* Prentice-Hall, Englewood Cliffs, New Jersey.

Turin, G.L. 1980. "Introduction to Spread Spectrum Antimultipath Techniques and Their Application to Urban Digital Radio," *IEEE Proceedings* 68(3):328–353.

Ungerboeck, G. 1982. "Channel Coding with Multilevel Phase Signals," *IEEE Transactions on Information Theory* 28(1):55–67.

Van Tuyl, R.L. 1996. "Unlicensed Millimeter Wave Communications: A New Opportunity for MMIC Technology at 60 GHz," paper presented at IEEE GaAs IC Symposium, Palo Alto, California, November 3–6.

Viterbi, A.J. 1994. "A Vision of the Second Century of Wireless Communication," *International Journal of Wireless Information Networks* 1(1):3–6.

Walrand, J., and P. Varaiya. 1996. *High Performance Communications Networks.* Morgan Kaufmann, San Francisco, California.

Wepman, J.A., and J.R. Hoffman. 1996. *RF and IF Digitization in Radio Receivers: Theory, Concepts, and Examples.* NTIA Report 96-328. U.S. Department of Commerce, Washington, D.C.

Wickelgren, I.J. 1996. "Local Area Networks Go Wireless," *IEEE Spectrum* 33(9):34–40.

Yacoub, M.D. 1993. *Foundations of Mobile Radio Engineering.* CRC Press, Boca Raton, Florida.

Ziemer, R.E., and W.H. Tranter. 1995. *Principles of Communications Systems, Modulation and Noise.* John Wiley and Sons, New York.

Ziv, J., and A. Lempel. 1978. "Compression of Individual Sequences by Variable Rate Source Coding," *IEEE Transactions on Information Theory* IT-24(5):530–536.

APPENDIXES

A

Biographies of Committee Members

DAVID J. GOODMAN, *chair*, is director of the Wireless Information Network Laboratory at Rutgers University, where he is also a professor and former chairman in the Department of Electrical and Computer Engineering. Previously, he spent 20 years at AT&T Bell Laboratories, where he was a department head in communications systems research. Dr. Goodman received a B.S. degree from Rensselaer Polytechnic Institute, an M.S. degree from New York University, and a Ph.D. from Imperial College, University of London, all in electrical engineering. He is a fellow of the Institute of Electrical and Electronics Engineers and the Institution of Electrical Engineers.

NORMAN ABRAMSON is vice president and chief technology officer of ALOHA Networks, Inc. Previously, he was director of the ALOHA System at the University of Hawaii and a professor of electrical engineering at Hawaii and at Stanford University. Dr. Abramson has taught communication theory, computer networks, and satellite communication courses at the University of California at Berkeley, Harvard University, and Massachusetts Institute of Technology while on visiting appointments. He is a recipient of the Koji Kobayashi Computers and Communications award given by the Institute of Electrical and Electronics Engineers. Dr. Abramson received an A.B. degree from Harvard University and an M.A. degree from the University of California at Los Angeles, both in physics, and a Ph.D. in electrical engineering from Stanford University.

EUGENE CACCIAMANI is a senior vice president at Hughes Network Systems, where he is responsible for business development, advanced systems engineering, and government business. Previously, he was president and chief executive officer of MA/COMNET and held key management positions at the American Satellite Company and the Communications Satellite Corporation. Dr. Cacciamani also worked at RCA Data Systems Division and served in the U.S. Air Force. He is a fellow of the Institute of Electrical and Electronics Engineers. He received a B.S. degree from Union College and M.S. and Ph.D. degrees from the Catholic University of America, all in electrical engineering.

JOEL ENGEL, recently retired as vice president-technology at Ameritech. Previously, he was vice president for research and development at MCI and vice president of engineering at Satellite Business Systems. Dr. Engel also held several positions, including manager for corporate planning studies, at AT&T and Bell Laboratories and at the Massachusetts Institute of Technology (MIT) Instrumentation Laboratory. He is a member of the National Academy of Engineering and a fellow of the Institute of Electrical and Electronics Engineers (IEEE). He is a recipient of the National Medal of Technology and the IEEE Alexander Graham Bell Medal. Dr. Engel received a B.S. degree from the City College of New York, an M.S. degree from MIT, and a Ph.D. degree from the Polytechnic Institute of Brooklyn.

MARK EPSTEIN is vice president, development, at QUALCOMM, Inc. Prior to joining QUALCOMM, Dr. Epstein served as deputy for C^3I to the assistant secretary of the U.S. Army (RDA), and, earlier, as a staff assistant in the Office of the Secretary of Defense. His previous positions include program director at the Computer Sciences Corporation and assistant director, engineering, for Northrop Page Communications. Dr. Epstein received B.S. and M.S. degrees from the Massachusetts Institute of Technology and a Ph.D. from Stanford University, all in electrical engineering. He serves as chairman of the Telecommunications Industry Association's International Standards Coordinating Committee and as a member of two U.S. delegations to the International Telecommunications Union Radiocommunication Sector.

BRUCE FETTE is chief engineer at Motorola's Government and Systems Technology Group Communications Division. He has been involved in signal processing analysis for numerous systems, including the

SpeakEASY multiband, multimode radio; wideband wireless network; Joint Special Forces Operations Radio System, Integrated Team Radio; and numerous vocoders and systems in the STU-III family of secure telephone systems. Dr. Fette received a B.S. degree from the University of Cincinnati and M.S. and Ph.D. degrees from Arizona State University, all in electrical engineering.

DOUGLAS C. FIELDS is a vice president of United Parcel Service, where he has been involved in developing wireless voice and data communications systems for the worldwide delivery vehicle fleet. Previously, he directed telecommunications planning for Corning Glass Works, Levi Strauss & Company, and Key Services Corporation and was a consultant to the Pacific Telephone Company. Mr. Fields attended Los Angeles City College and served in the U.S. Army Signal Corps. He is a member of the International Telecommunications Users Group.

BEZALEL GAVISH is a professor at the Owen Graduate School of Management at Vanderbilt University. He previously held positions at the Norwegian School of Economics and Business Administration, Naval Postgraduate School, Technion-Israel Institute of Technology, and University of Rochester. In addition, he was a visiting researcher at Bell Laboratories and IBM Corp. and a department head for the Israel Defense Forces logistics branch. Dr. Gavish received a B.Sc. degree in industrial engineering and management science and M.Sc. and Ph.D. degrees in operations research, all from Technion.

ANDREA GOLDSMITH is an assistant professor of electrical engineering at the California Institute of Technology. She previously worked at AT&T Bell Laboratories and MAXIM Technologies. Dr. Goldsmith is the recipient of a National Science Foundation career development award and was an IBM fellow. She is an editor of both *IEEE Transactions on Communications* and *IEEE Personal Communications Magazine*. Dr. Goldsmith received B.S., M.S., and Ph.D. degrees from the University of California at Berkeley, all in electrical engineering.

RANDY H. KATZ is a professor of computer science at the University of California at Berkeley and a principal investigator in the Bay Area Wireless Access Network project. He has taught at UC Berkeley since 1983, with the exception of 1993 and 1994, when he was a program manager and deputy director of the Computing Systems Technology Office at the Defense Advanced Research Projects Agency. Dr. Katz received a B.S. degree from Cornell University and M.S. and Ph.D. degrees from UC Berkeley, all in computer science. He is a fellow of the Association for Computing Machinery and of the Institute of Electrical and Electronics Engineers.

EDWIN A. KELLEY is division chief scientist and manager of the digital transceiver product line for both commercial and military applications at Hughes Communication Products, Hughes Aircraft Company. Mr. Kelley has been the program manager for the advanced secure digital radio and the advanced communications engine project supported by the Department of Defense. Mr. Kelley received a B.S. degree in electrical engineering and M.S. degrees in electrical engineering and computer science, all from the University of California at Berkeley.

KAVEH PAHLAVAN is director of the Center for Wireless Information Network Studies at the Worcester Polytechnic Institute, where he is also a professor. Previously, he was the director of advanced development at Infinit, Inc., and taught at Northeastern University. He is a consultant to numerous companies on cellular networks, communications theory, and technical aspects of wireless networks. Dr. Pahlavan is a fellow of the Institute of Electrical and Electronics Engineers Communication Society and the editor-in-chief of the *International Journal of Wireless Information Networks.* He received an M.S. degree from the University of Tehran and a Ph.D. degree from Worcester Polytechnic Institute, both in electrical engineering.

CHARLES E. PERKINS is a senior staff engineer at Sun Microsystems. Previously, he was a research staff member at the IBM T.J. Watson Research Center. He is the author or co-author of standards-track documents in the srvloc (service location), dhc (dynamic host configuration), and IPng (IP new generation) working groups of the Internet Engineering Task Force. He also serves on the Internet Architecture Board and recently wrote a book on Mobile IP. Mr. Perkins is an associate editor of *Mobile Communications and Computing Review*, the official publication of ACM SIGMOBILE, and serves on the editorial board of *IEEE/ACM Transactions on Networking*. He received a B.A. degree in mathematics and an M.E.E. degree from Rice University, and an M.A. degree in mathematics from Columbia University.

THEODORE RAPPAPORT is professor and director of the Mobile and Portable Radio Research Group in the Bradley Department of Electrical Engineering at Virginia Polytechnic Institute. Previously, he was a postdoctoral research associate at the National Science Foundation (NSF) Engineering Research Center for Intelligent Manufacturing Systems at Purdue University and an engineer at Harris Corp. Dr.

Rappaport is a fellow of the Radio Club of America and a recipient of the NSF presidential faculty fellow award and the Institute of Electrical and Electronics Engineers Marconi Young Scientist Award. He received B.S., M.S., and Ph.D. degrees from Purdue University, all in electrical engineering.

JESSE RUSSELL, is chief wireless architect and managing director of the AT&T Wireless Communications Center of Excellence. His previous positions at AT&T include chief technical officer in the Wireless Systems Business Unit, vice president of the Advanced Wireless Technology Laboratory, director of the Cellular Transmission Laboratory and Cellular Telecommunications Laboratory, and head of several departments. Mr. Russell is a member of the National Academy of Engineering and a fellow of the Institute of Electrical and Electronics Engineers. He has received numerous awards, including U.S. Black Engineer of the Year. He received a B.S. degree from Tennessee State University and an M.S. from Stanford University, both in electrical engineering.

B

Briefers to the Committee

JULY 29, 1996

Kevin Mills, Program Manager, Defense Advanced Research Projects Agency

Eugene Famolari, Associate Director, Army Communications-Electronics Command (CECOM)

Paul Sass, Special Projects Office, CECOM

Jim Freebersyser, Army Research Office

John Graniero, Chief Scientist for C3, Air Force Rome Laboratory

OCTOBER 15, 1996

Rob Ruth, Program Manager, Defense Advanced Research Projects Agency

Marlan Kvigne, Chief Engineer, Communications and Information Systems Department, NCCOSC (RDTE Division)

Joe Macker, Network Research Scientist, Center for High Assurance Computer Systems, Naval Research Laboratory

David Wye, Technical Advisor, Wireless Telecommunications Bureau, Federal Communications Commission

DECEMBER 10, 1996

Lou Dellaverson, Manager, Wireless ATM Laboratory, Motorola

Nicolas Kauser, Chief Technology Officer, AT&T Wireless Services
William Osborn, Manager of Technology Development, Cellular Phone Research and Development Center, Ericsson, Inc.
Joseph A. Tarallo, Director, Wireless Base Station and Radio Technology Department, Lucent Technologies
D. Raychaudhuri, NEC America (via telephone)

At the December 1996 session, invited speakers from industry were asked to address these questions, among others:

1. What technologies will the consumer wireless industry likely develop over the next 5 to 15 years, regardless of whether the federal government provides basic R&D?
2. Are there any critical telecommunications technologies that must be funded/developed by the U.S. government because commercial industry cannot justify the risk or exploratory expense?
3. What are the most critical technical and nontechnical issues facing the wireless industry that threaten the competitiveness or growth of individual companies on a global scale?
4. What is the potential for synergy between military needs and likely commercial development? What technology gaps will the military need to fill in order to use commercial products and services?
5. Which countries are leading in wireless communications, in terms of deployment of technologies? What are examples of good policies that help to foster good technological development in these countries? Who will be the future leaders, and why?
6. How does industry benefit from wireless research done in academic institutions, universities, and research centers? What are some examples?
7. Given that the military may have to operate globally in developed and underdeveloped regions, what could be viable wireless technologies to support military mobile missions?

In addition to the guests invited to participate on the panel, there were seven observers from the Federal Communications Commission: David Wye, Ron Netro, Marty Liebman, Mike Marcus, Larry Petak, Steve Sharkey, and Tom Stanley.

C

Glossary

ACE (advanced communications engine) is a prototype multiband, multimode software radio capable of emulating both military combat net radios and commercial avionics radio systems.

ACN (airborne communications node) is an unmanned aerial vehicle designed and equipped to provide hierarchical communications services and crosslinking over a broad theater of operations. See RAP and UAV.

ACTS (Advanced Communications Technologies and Services program) is the European Commission's latest precompetitive research effort focusing on advanced communications systems.

A/D converter is an analog-to-digital converter.

AJ (antijam) refers to techniques for reducing the effectiveness of attempts to jam communications channels. See Jamming.

AMPS (advanced mobile phone system) is the standard for the analog cellular radio telephones now widely available throughout the United States.

ARQ (automatic repeat request) is a protocol used to retransmit data packets received in error.

ASIC is an application-specific integrated circuit.

ATM (asynchronous transfer mode) enables voice, data, and video to be handled with a uniform transmission protocol. It breaks up the information to be transmitted into short packets of data and intersperses them using time division with data from other sources and delivered over trunk networks.

ATM cell is an information packet containing 53 bytes of data traffic and additional "overhead" defining the virtual circuits and paths over which the data are to be transmitted.

Bandwidth is the part of a frequency band occupied by a communications channel. Sometimes the term is used to describe the number of bits per second transmitted in a channel.

Bent pipe is a satellite communications system that transmits to an Earth station essentially the same signal it receives from another Earth station.

BER (bit-error rate) is the probability that a bit is received in error.

Bit is a binary unit of information.

bps (bits per second) refers to the speed at which data is generated by a source or transmitted over a communications channel. Measurements are often stated in units of 10^3 bits per second (kilobits or kbps) or 10^6 bits per second (megabits or Mbps).

Byte is a unit of 8 bits.

C^{4}I (command, control, communications, computing, intelligence) is a military concept encompassing all the functions and capabilities sought in an advanced communications system.

CDMA (code division multiple access) is a technique that allows many users to share the same radio spectrum. A sequence of pseudo-random bits, known as a code, spreads the information signal over a much larger range of frequencies than is occupied by the original information signal. See FDMA, TDMA, spread spectrum.

CDPD (cellular digital packet data) is a packet-switched network that uses one or more channels in an analog cellular telephone system.

Cell refers to a geographic region within which cellular telephone subscribers can communicate with a particular base station (site). Cell radius ranges from 0.5–15 kilometers depending on the density of the subscribers and the extent of topological obstructions.

CELP (code excited linear prediction) is a technique for encoding voice.

CMOS (complementary metal oxide semiconductor) is an inexpensive, low-power integrated circuit technology.

CONDOR is a National Security Agency program designed to develop and demonstrate secure voice and secure net broadcast services, using STU III-compatible units, over the commercial cellular infrastructure.

COTS (commercial off-the-shelf) refers to readily available commercial technologies and systems.

CSMA/CD (carrier-sense multiple access with collision detection) is a protocol that regulates the manner in which terminals gain access to a shared communications channel.

DAMA (demand-assigned multiple access) is a technique that enables many users to share the same radio spectrum. A common signaling channel is assigned to handle requests from transmitters for network capacity.

dB is decibels, a unit for expressing the relative intensity of acoustic and electromagnetic waves.

DBS (direct broadcast satellite) is a system in which GEO satellites broadcast a signal with sufficient power to enable direct reception in a home, office, or vehicle with an inexpensive receiver.

DFE (decision feedback equalizer) is a nonlinear equalization technique designed to reduce the effects of intersymbol interference. See ISI.

Doppler effect is a change in the received signal frequency due to movement of the transmitter or receiver.

DSP (digital signal processor/processing) is a specialized integrated circuit used to analyze or alter the characteristics of communications signals.

Erlang is a unit reflecting the traffic intensity on a communication link. It is equivalent to the fraction of time that the link is occupied.

ETSI is the European Telecommunications Standards Institute.

Fading refers to changes in the amplitude of received signals due to characteristics of the transmission path and motion of the transmitter and receiver.

FCC is the Federal Communications Commission.

FDMA (frequency division multiple access) is a technique that enables many users to share the same radio spectrum. Each user is allocated a different frequency. This is the approach used in AMPS cellular radio. See CDMA, TDMA.

FEC (forward error correction) codes give digital signals a highly specialized redundancy that enables a receiver to recognize and correct occasional errors in the received signal.

FH (frequency hopping) is a technique for changing transmit frequency in a way that makes it difficult for an adversary to jam communications.

FIR (finite impulse response) refers to a signal-processing operation performed by a DSP.

FM (frequency modulation) is a modulation technique in which information is conveyed in the high frequency of a carrier signal.

FPLMTS (future public land mobile telecommunication system) is the original name for the International Telecommunications Union's concept for third-generation wireless telephone systems. It is now called IMT-2000.

Frequency reuse refers to the use of the same signal spectrum at different geographical locations.

GBS (global broadcast system) is an advanced military satellite communications system designed to have very high data rates (100 megabits per second) and global coverage.

GEO (geosynchronous orbit) refers to an equatorial satellite orbit approximately 36,000 kilometers from the Earth in which the satellite remains stationary over one position on the Earth's surface.

GloMo (Global Mobile Information Systems) is a research and demonstration program dealing with mobile, wireless communications. It is funded by the Defense Advanced Research Projects Agency.

GSM (global system for mobile communications) is a combination TDMA/FDMA cellular radio system that is the current digital standard in Europe. It was originally known as group system mobile.

Have Quick is a military UHF radio designed to provide secure air-to-air and air-to-ground communications with AJ capabilities.

HF (high frequency) is the frequency band at 3–30 megahertz.

HIPERLAN is a high-performance radio local-area network designed to operate at 20 megabits per second.

Hz (hertz) refers to a unit of frequency equal to one cycle per second. Frequencies are often stated in units of 10^3 hertz (kilohertz or kHz), 10^6 hertz (megahertz or MHz), or 10^9 hertz (gigahertz or GHz).

IC is an integrated circuit.

IEEE is the Institute of Electrical and Electronics Engineers.

IMT-2000 (International Mobile Telecommunications-2000) is the International Telecommunications Union's concept for third-generation mobile wireless telephone systems.

Information warfare refers primarily to recent U.S. initiatives designed to protect computer network infrastructures against intentional disruptions. The term encompasses many forms of disruption aimed at communications networks (both wired and wireless) and the relevant countermeasures.

INMARSAT (International Maritime Satellite) is an organization of 75 member countries that has launched several generations of satellite communications systems for voice and low-rate data applications.

Intelsat is an international, government-chartered organization established in 1964 to coordinate worldwide satellite communications programs. A series of INTELSAT communications satellites has been launched.

IS-95 is the North American standard for second-generation CDMA digital wireless telephone systems.

IS-136 is the North American standard for second-generation TDMA digital wireless telephone systems.

ISDN (integrated services digital network) is a set of international standards that specify the manner in which different types of information (e.g., voice, data, video) can be transmitted in the same communications system.

ISI (intersymbol interference) occurs when multipath reflections corresponding to a given bit transmission arrive at the receiver simultaneously with subsequent data bits.

ISM (industrial, scientific, and medical) refers to the unlicensed frequency bands available for use by any wireless device that conforms to rules established by the FCC.

ITU is the International Telecommunications Union.

Jamming is a type of transmission designed to disrupt the radio communications of an adversary so as to interfere with military operations.

JCIT (Joint C^4I Terminal) is a software-defined radio under development by the Naval Research Laboratory that will implement combat net, intelligence communications, and military datalinks on a single platform.

LAN is a local area network.

LEO (low Earth orbit) is a satellite communications system deployed in a low orbit (500 to 2,000 kilometers from Earth).

LOS (line of sight) refers to a radio communications systems with an unobstructed path between the transmitter and receiver.

LPC (linear predictive coding) is a coding technique based on a mathematical model of a voice signal.

LPD/I (low probability of detection and interception) refers to the capability to minimize an adversary's awareness of transmitted radio energy, ability to measure any properties of a detected signal, or ability to intercept the transmission.

MEO (medium Earth orbit) is a satellite orbit of intermediate height between LEO and GEO orbits.

Message store-forward is a system in which a message is uploaded to a satellite and held there until the satellite is in a position to download it to the destination. It is an alternative to bent-pipe systems.

Millennium is a military research effort to design an ultra-wideband radio. One objective is to demonstrate extremely high-speed A/D converters for military and commercial applications.

MISSI (Multilevel Information Systems Security Initiative) is a National Security Agency initiative designed to provide a framework for the development of interoperable, complementary security products.

MMITS (Modular Multifunction Information Transfer System) is an industry forum focusing on technical standards and applications for software-defined radios, networking radios, and multimode radios.

Mobile IP (mobile internetworking routing protocol) is designed to support mobile Internet users.

MSE (mobile subscriber equipment) is a military radio that resembles a cellular telephone.

MSRT (mobile subscriber radio terminal) is a military radio.

Multipath propagation is a phenomenon in which copies of a transmitted signal arrive at different times at a receiver.

NATO is the North Atlantic Treaty Organization.

NES (network encryption system) is an encryption system certified by the National Security Agency that enables clusters of defense computer networks to interconnect through the unclassified Internet.

OSI (Open Systems Interconnection) refers to a model that describes networks as a series of layers.

Packet is a collection of information symbols transmitted together on a communications channel.

PACS (personal access communications system) is a U.S. standard for second-generation digital wireless telephone systems serving subscribers moving at pedestrian speeds.

PDA (personal digital assistant) is a portable, low-power computing device with a small display used for information storage.

Peer-to-peer is a network architecture in which transmissions flow between terminals without passing through a central hub.

Personal communications services are offered in frequency bands around 1900 megahertz allocated for this purpose by the FCC.

PHS (personal handyphone system) is a Japanese standard for mobile digital telephone services.

PSTN (public switched telephone network) is a collection of worldwide wired telephone networks.

QoS (quality of service) refers to end-to-end performance guarantees offered by a network.

RACE (Research for Advanced Communications in Europe) was a program funded by the European Commission to perform precompetitive research on advanced communications systems. It was superseded by the ACTS program.

Radio spectrum is a range of radio frequencies required to support one or more communications channels.

RAM is random-access memory.

RAP (radio access point) is a wheeled or tracked vehicle that carries extensive communication systems, including an "on the move" antenna system. It provides a variety of cross-networking, repeater, and information services.

RF is radio frequency.

RMS (root mean square) is the standard deviation of a random variable.

RSVP (resource reservation protocol) supports the delivery of real-time information over the Internet.

RTP (real-time protocol) is designed to support delay-intolerant data streams, such as video, transmitted over the Internet.

S3 (Scalable Self-Organizing Simulations) is a research program that uses parallel computers to simulate communications networks. The program is sponsored by the Defense Advanced Research Projects Agency and the National Science Foundation.

SHF (superhigh frequency) is the frequency band at 3–30 gigahertz used for satellite, radar, and microwave communications.

SINCGARS (single-channel ground and airborne radio system) is a military radio that hops transmission frequencies within the very-high-frequency band, a capability that helps prevent jamming.

SNR (signal-to-noise ratio) is the ratio of signal power to noise power. The higher the SNR, the clearer the transmission.

Software-defined radio is a radio implemented on DSPs with functions defined by software, which can be downloaded as needed. Such radios can use many types of modulations.

SpeakEASY is a software-defined radio designed and demonstrated with support from the Defense Advanced Research Projects Agency. SpeakEASY can emulate some legacy military radio systems and operate across a wide frequency range.

Spectral link efficiency is the data rate (in bits per second) per unit of bandwidth of a communications channel.

Spread spectrum is a technique in which a signal is spread over a much larger frequency range than the minimum required to deliver the message. It is often used by the military for antijam purposes. It is used by the commercial sector in CDMA mobile telephone systems and wireless LANs.